Unlocking Offshore Wind: The Efficiency and Flexibility of HVDC Transmission

Jacksin

Contents

1. Introduction

1.1. HVDC Transmission

Traditionally, high-voltage direct current (HVDC) transmission has been used for efficient transport of bulk power over long distance [EPR]. This is because, compared to HVAC transmission, HVDC offers several technical as well as economic advantages such as lower investment costs, lower losses and smaller right-of-way for the same power transmission over a long distance [PBHK17]. In addition, HVDC enables interconnection of asynchronous AC systems including systems operating at different frequencies (e.g., 50 Hz and 60 Hz) and offers flexible and efficient control, which makes it suitable for integration of (intermittent) renewable energy sources. Moreover, HVDC has no limitation on the transmission distance via underground and submarine cables offering flexibility in choice of substation location. For example, in the UK and Germany several offshore HVDC links could have been AC, but were implemented as HVDC to allow locating the onshore substation at a location where it would have lower environmental impact and have a better grid connection.

Today, locating wind farms offshore has become appealing since wind speed tends to be higher on average and also more stable. However, as the distance from shore increases, evacuation of power from such remote locations becomes challenging using the traditional HVAC transmission. This is because long HVAC cables require reactive power compensation for efficient transport of active power. For such applications HVDC transmission proves to be a viable solution. As a result, building point-to-point HVDC connections has become an integral part of power transmission from remote offshore wind farms to the onshore grid. Nowadays, several point-to-point offshore HVDC links are in operation, connecting large offshore wind resources in the North Sea to the European mainland, and many are under construction [ISBB16].

The other drivers for HVDC in Europe are market integration as evidenced by the growth of interconnectors – targeting 15% interconnection by 2030, power from shore to oil and gas platforms, and reinforcement of onshore AC grids.

1.2. HVDC Converter Technologies

HVDC technology has evolved through several technological stages – from mercury arc valves to thyristor valves based line commutated converters (LCC) and to the voltage source converters (VSC). Compared to the LCC technology, the VSC technology offers several new capabilities and features such as bi-directional power flow without voltage polarity reversal – making it ideally suited for multi-terminal HVDC (MTDC) grid development, independent control of active and reactive power – a vital feature for connection to weak or even passive grids, smaller footprint – essentially suitable for offshore application, etc. The VSC technology itself has evolved significantly, from a simple two level converter to the state-of-the-art multi-level converters.

In fact both converter technologies, the LCC (also known as current source converter (CSC)) and the VSC, are in use today. The choice between the two depends on many factors including the transmission power and the voltage rating, losses, the strength and behavior of the connected AC grid, etc.

1.2.1. Line commutated/current source converter (LCC/CSC)

LCC is a mature technology that has been exploited for more than 60 years in several projects worldwide. The basic form of the converter is the six-pulse Graetz-bridge, shown in Figure 1.1. Each of the arms (numbered T1-T6 in Figure 1.1) consists of thyristor valves, each of which, in turn, is composed of a large number of thyristors connected in series to attain the required blocking capability as shown in the figure.

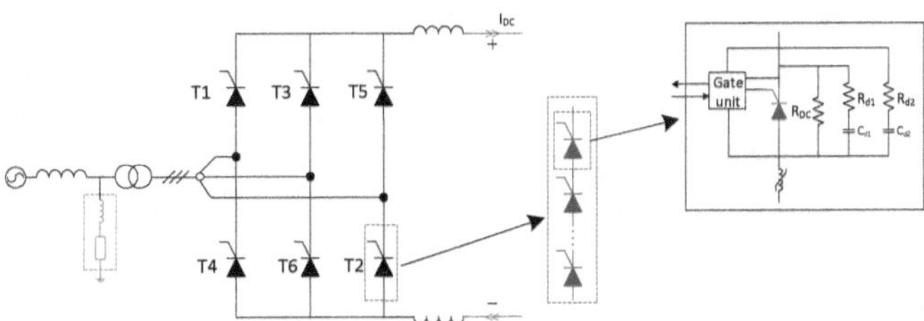

Figure 1.1.: Three-phase Graetz bridge – six-pulse LCC topology based on thyristors [EPR]

A thyristor is a semi-controllable power electronic (PE) device which can be turned-on, but cannot be turned-off by a control command. It is naturally turned-off when the current flow direction changes by an external circuit – thus, a converter made of such devices is named line (voltage) commutated converter.

Today, LCC technology is very much suited for bulk power transmission over a long distance. Power transmission up to 12 GW at ultra-high voltage of 1100 kV has been achieved [Gou19].

1.2.2. Voltage source converters (VSC)

The VSC is based on fully controllable PE switches such as insulated gate bipolar transistors (IGBTs) and is able to control the output voltage magnitude, phase angle and frequency [SHN$^+$16]. Thus, the VSC enables independent control of active and reactive power flow between the converter and the AC grid.

Two topologies of the VSC exist; two-level and multi-level converters. The two-level converter topology has the same circuit layout as the basic LCC – the Graetz-bridge. However, instead of thyristors, the converter valves are composed of series connected IGBTs[1]. Two-level VSC topology is widely used at low voltage below 1800 V [SHN$^+$16]. At high voltage, however, it is restricted in capacity to around 400 MW and losses are considerably high at approximately 2 - 3% per converter [SHN$^+$16, BPST].

A number of multi-level VSC topologies has been developed to reduce switching losses and harmonics generated on one hand, and increase power transmission capacity on the other hand. Multi-level converters, particularly the modular multi-level converters (MMCs), have improved on these key areas and have come to dominate the market for new installations [SHN$^+$16].

1.2.2.1. Modular multi-level converter (MMC)

A typical MMC is shown in Figure 1.2a and is composed of the basic building blocks called sub-modules (SMs), which are shown in Figure 1.2b and 1.2c. MMC can be built from either half-bridge (Figure 1.2b) or full-bridge (Figure 1.2c) SMs, each construction having associated pros and cons. Compared to a converter based on half-bridge SMs, a converter built from full-bridge SMs has DC fault current blocking capability; however, this comes at the price of twice the number of PE semiconductors, which results in increased cost and operational loss.

The modular approach using a large number of SMs allows high power throughput at reduced power loss. Moreover, with a greater number of output voltage levels, harmonics generated from switching decreases and, in most cases, filtering is not required. In general, scalability, modularity, enhanced reliability and high-efficiency makes MMC an

---

[1] each IGBT having a freewheeling anti-parallel diode

appropriate choice for high-power applications, and nowadays the VSC based HVDC transmissions are exclusively based on this technology.

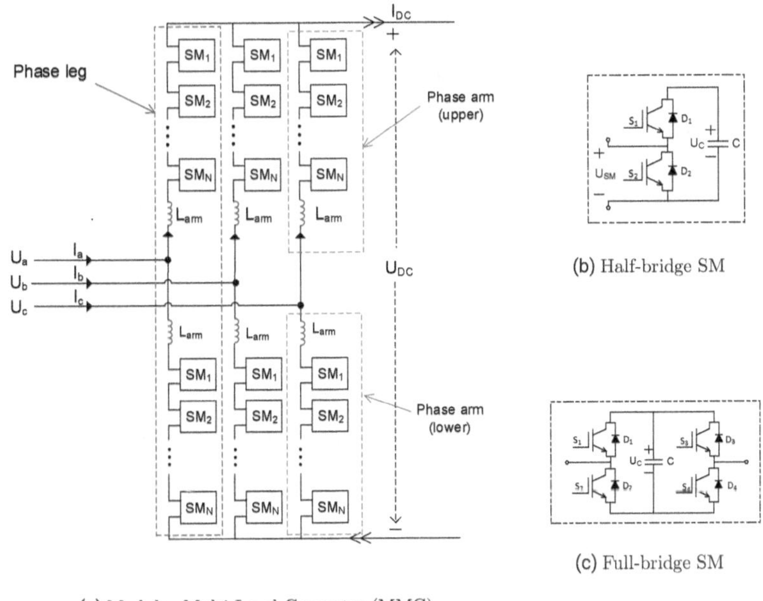

(b) Half-bridge SM

(c) Full-bridge SM

(a) Modular Multi-Level Converter (MMC)

Figure 1.2.: Modular Multi-level Converter

1.3. Why Multi-Terminal HVDC Grid? What are the Challenges?

The same technical and economic benefits such as flexibility, reliability and resource sharing obtained from the present day interconnected AC system can be obtained from building multi-terminal, meshed HVDC grids; in fact, with many additional advantages of HVDC power transmission described above.

Today several point-to-point HVDC systems are in operation, many projects are under construction or in the planning phase – see Figure 1.3. For example, considering Europe,

more than 50 HVDC systems are in operation[2] and more than 15 are under construction
with up to 11 projects expected to be commissioned until 2030 [Wik, PRO]. At least
25 HVDC point-to-point systems are in operation in the North Sea by 2020 [ISBB16].
Moreover, the European Union (EU) aims to be climate-neutral by 2050 – an economy
with net-zero greenhouse gas emissions. In order to meet this target, the EU envisions
about 300 GW power generation from offshore wind by 2050, of which more than 210 GW
is expected to be generated in the North Sea [FFH+]. Most of the generated power will
be transported to shore as electricity mainly via HVDC transmission.

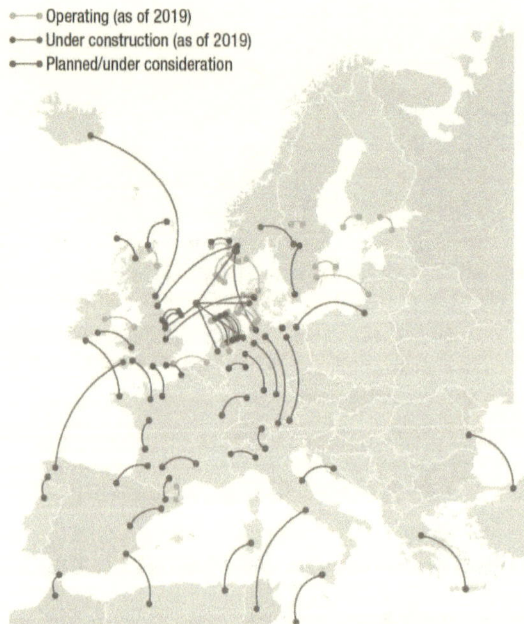

Figure 1.3.: Construction of VSC HVDC in Europe [NAO]

---

[2]The list includes LCC projects.

On the other hand, several studies show that MTDC grids are viable solution for large scale integration of renewable energy sources such as offshore wind farms [LZW+14, XY11, PBR+13, CLBR12, YFO10, EHB12]. Hence, instead of building multiple point-to-point HVDC connections, development of MTDC grids seems to be the next logical target [BPST, ISBB16, SOKL20]. In such a case, MTDC grids can improve security of supply by smoothing power output, sharing capacity and interconnecting various national networks.

In fact, in most parts of the world, MTDC grids may not be built anew from scratch although this is happening in China where three MTDC pilot projects are in operation [Rao15, THP+15, ATW17]. The most organic approach is to build MTDC grids gradually by interconnecting existing systems. There are several technical challenges of building MTDC grids based on the already operating systems. For example, most of these systems are operating at different voltage levels – ranging from 80 kV up to 600 kV. To interconnect such systems DC/DC converters are required. The latter are not mature yet for high-voltage applications although recently MMC based DC/DC converters are gaining a lot of attention [ATW17, PFM+19].

Perhaps, a more organic approach to the development of MTDC grids is first to interconnect systems with the same transmission voltage – for example by interconnecting HVDC systems operating at ±320 kV. Also in this case there are challenges; the components, the control and protection functions of the existing systems are optimized for point-to-point power transmission. Furthermore, in many projects different converter configurations (symmetric monopole, (rigid or metallic return) bipole and asymmetric monopole) are used while it is needless to mention the interoperability issues arising from intellectual property (IP) related matters associated with different vendors. The lack of standardization and absence of clearly defined and harmonized DC grid code requirements are also some of the hurdles that need to be tackled before building MTDC grids. Moreover, as the complexity of MTDC increases, dedicated power flow controllers are required for optimal power transmission in such a grid [HS20].

Nevertheless, the major technical challenge hindering the deployment of MTDC grids is the lack of a well proven protection system that is able to clear DC side faults without leading to the collapse of the entire DC network. At the heart of such a challenge is the lack of fast, low loss and reliable HVDC circuit breakers (CBs) capable of clearing DC faults [Fra11].

1.3.1. Motivation and background

Since the inception of HVDC transmission in the 1950's, there has always been a plan to build MTDC grids; and hence, the HVDC CBs have been the subject of significant research. Notably, from the 1970s until the late 1990s substantial research and development has been conducted [YTI$^+$82, SKT$^+$81, AYT$^+$85, NNH$^+$01, GBK72, HLRS76, STH$^+$84, LSY$^+$85, TSK$^+$80, BHK74, LH74, HLK73, AC78, Pre82, VCP$^+$85, KWK$^+$86, BMR$^+$85, THYK87, PMR$^+$88, IHI$^+$97]. AC CBs with auxiliary circuits for current zero creation and energy absorption have been thoroughly explored.

Nevertheless, many factors hindered further development and application of HVDC CBs in the 1980s and 90s. Some of these are:

1. The HVDC CBs were aimed for the protection of MTDC grids built from the LCC technology. The rate-of-rise and the magnitude of fault currents in such systems are not as critical as in the MTDC grids built from the VSC HVDC – today's preferred converter technology for MTDC grids. The former systems naturally have large DC side smoothing reactors, which significantly reduce the rate-of-rise of fault current.

2. The view that the LCC HVDC converters have sophisticated and fast responding controls, which can be utilized to reduce the DC voltage and current in the event of a short-circuit. For example, the system fault current can be limited with the control of firing angles of the thyristor valves ("forced retardation") [EHL$^+$76, YTI$^+$82, EPR], whereas the currently envisaged MTDC grids operate in an analogous fashion as to the AC systems where the faulty section needs to be taken out of service with a minimum disturbance to the operation of the healthy part of the system.

3. The economic feasibility of HVDC CBs has been the subject of discussion – even to this date. For example, when the VI based HVDC CB was introduced for the first time in [GL72], it was questioned whether such a solution was economically sound, although the then new development could achieve much higher current interruption at much shorter duration (5-10 ms) [DBIP79]. Oftentimes, the cost and hence the price as well as the size and the complexity of an HVDC CB is judged in comparison with an AC CB of similar rating.

Therefore, the primary focus of HVDC CB designers was on the method of creating current zero(s) through the interruption unit(s) while the time to interrupt the fault current was less critical. Comparing the incremental benefit obtained by increasing the fault current clearing time against the extra cost needed for achieving the benefit was one of the challenges [PMR$^+$88].

Also, the requirements of the test circuits used for the performance verification of those breakers were defined accordingly. The test circuits are designed to supply (quasi-) DC current of required magnitude for a relatively longer duration, while the rate-of-rise of current and more importantly the magnitude of the driving voltage were less essential.

However, with the advent of the VSC HVDC technology, new challenges arose. Especially, the meshed MTDC grids are likely built from MMC constituting half-bridge SMs –

see Figure 1.2. This is because of the high investment cost and losses associated with other topologies such as full-bridge based MMC [ANN+12]. Also, the impact on AC grids due to non-selective fault clearing in case of full-bridge converters is often thought to be too high. However, when a fault occurs on the DC side of MTDC grids built from half-bridge converters, the anti-parallel diodes of the IGBTs in the SMs are forward biased and start to conduct and rectify AC fault current, essentially turning the converter into a diode rectifier. Besides, due to the capacitive behavior of HVDC cables and overall low DC side impedance, the fault current rises rapidly to a large magnitude that can subsequently damage system components if not rapidly handled [AAM+15]. Thus, in MTDC grids, extremely fast HVDC CBs enabling selective clearing of the faulty part without endangering the integrity of the system as a whole are necessary.

In fact the actual requirements for the HVDC CBs are, ultimately, determined by the functional specification of the HVDC system in which they are installed. Nevertheless, the progress in the development of the PE components has enhanced fault-ride-through capability of HVDC converters. In this regard, the HVDC CBs are expected to clear the DC line faults preferably before any of the converters or at most only one converter close to the fault location blocks[3] [BPS18b]. This, for example, is one of the criteria set forth for the Chinese Zhangbei project [PW18].

Following recent advances in PEs, a few industrial concepts of HVDC CBs, which can fulfill the stringent requirements, have been developed in the past few years [ZWZ+15, HJ11, GDPV14, EBH14]. Besides, the application of mechanical HVDC CBs based on the active current injection scheme has revived because of the developments in fast actuators; for instance, based on the electromagnetic drives [EBH14, TKT+14, BME15, TTKY08, BHÁ+19].

As a result, the test requirements of the latest technologies of HVDC CBs have also changed, although the performance demonstration of these concepts under actual short-circuit conditions in a MTDC grid is not conducted yet. These industrial concepts have been tested primarily for their speed and maximum current breaking capability using various test circuits as part of research and development at the respective laboratories. Mostly, a test circuit based on charged capacitor banks is used because of its simplicity and convenience [HJ11, GDPV14, ZWZ+15, TZH+, ZYZ+20]. The main challenge with such a test circuit is that large capacitor banks together with large inductors are required to produce sufficient stresses at rated values. AC short-circuit generators at power frequency (50 Hz) have been used for testing the recently developed active current injection HVDC CBs [EBH14, TKT+14, TSK+15]. In this case the breaker interrupts the AC current at

---

[3]stops its controlled operation because a turn-OFF signal is applied continuously to all IGBTs of the converter

its peak well before its natural zero crossing. This method lacks sufficient voltage stress especially during the energy absorption phase of the interruption process. Moreover, in none of the test circuits the application of DC recovery voltage was considered.

Sooner or later it becomes inevitable to demonstrate the performances of the HVDC CBs under actual operation conditions before installing them in service. Therefore, a test circuit capable of applying the required stresses (at rated system parameters) to the latest developments of HVDC CBs is essential. Currently, there is no test facility equipped with a sufficiently high-power DC source capable of testing HVDC CB. The optimal test circuit is indeed a full-scale converter connected to short-circuit generators having sufficient short-circuit power. However, the investment cost needed to build one at a test facility is prohibitive. In addition, due to the current and near future market potential for HVDC CBs, it makes sense to look for alternative but adequate test methods and circuits.

1.4. Scope and Objectives of the Work

1.4.1. Scope

At the moment of writing this book, no international standards specifying the test requirements of HVDC CBs exist. Generic test requirements have been derived from system studies, which can be broadly classified into four c ategories: dielectric, operational, breaking and endurance tests. Among these tests, short-circuit current breaking/interruption tests are the most challenging. The remaining tests are not special in that they can be covered by standard circuits and procedures at test laboratories. The latter will not be considered in this book since there are no special requirements nor special test circuits requiring in-depth research.

Adequate test circuits are essential for rated short-circuit current breaking performance verification of H VDC C Bs. I deally, t he t esting o f H VDC C Bs is d one t hrough t he use of a high-power DC source – supplying high-current and high-voltage simultaneously – which is not available at any test laboratory worldwide. Given the current stage of development and market potential it is unlikely that expensive and elaborate test circuits will be set up for solely testing HVDC CB prototypes. Thus, alternative test circuits providing equivalent stresses are urgently sought. Hence the focus of this book is on the development of a test method, design and implementation of an adequate test circuit that can be used for DC short-circuit current breaking tests.

Before designing a test circuit, the critical stages of a fault current interruption process and the actual stresses to which the HVDC CBs are subjected during the process must be understood. Thus, it is necessary to define test conditions and setup test requirements, before embarking on characterization of a suitable test circuit. For this purpose, simulation study of fault conditions in an MTDC grid is necessary. To define and refine the stresses to which HVDC CBs are subjected, fault current interruption by simulation models of HVDC CB technologies in a conceptual MTDC grid needs to be performed.

1.4.2. Objectives

Tests can be performed either to proof a concept (verify operation and performance as well as explore limits) or to confirm the rating and desired functionality of an HVDC CB. The latter is the focus of this book – developing an adequate test method, design and implementation of a test circuit to determine the rated short-circuit current interruption performance of HVDC CBs.

1.4.2.1. Major objective

The main objective of this book is, first, to define test requirements of HVDC CBs; then, to develop a test method based on available infrastructure at a test facility; and finally based on the developed test method, to design and implement an adequate test circuit in a test laboratory.

1.4.2.2. Specific objectives

Specifically the book is aiming at the following sub-objectives:

- Identify the critical challenges and define the generic requirements of DC short-circuit current interruption.

- Review and evaluate different concepts and realizations of HVDC CBs. Illustrate operation principles, review various concepts and discuss the recently developed prototypes of HVDC CBs together with performance achievements as reported in the literature.

- Investigate fault currents in a conceptual MTDC grid. Identify the temporal stages of fault current development, the main contributors to the fault current at each stage and the system parameters that have an impact on a fault current in a MTDC grid via EMT simulation studies.

- Develop EMT simulation models of the different concepts of HVDC CBs. Embed these models into a conceptual MTDC grid to identify the critical stages of the short-circuit current interruption process. Following this, identify and analyze the electrical stresses to which these breakers are subjected when interrupting fault currents. Translate the stresses into the test requirements of HVDC CBs, which must be reproduced during a test.

- Using the identified stresses as guidelines, define the requirements of HVDC CB test circuits.

- Evaluate and compare various test methods along with the associated test circuits with respect to the identified requirements while considering practical feasibility and/or availability.

- Propose a new test method based on existing infrastructure at a test facility. Identify the key parameters of the proposed test circuit and demonstrate feasibility in a test laboratory.

- Design and implement a complete test circuit based on the proposed method. Demonstrate the implemented test circuit which is capable of supplying all the critical stresses at each stage of the current interruption process by performing tests on actual HVDC CBs supplied by original equipment manufacturers (OEMs) within the project framework[4].

- Identify the challenges and the limitations associated with the developed test method while considering the available facilities at a test laboratory. Propose and demonstrate practical mitigation methods to address these challenges.

- Set up an experimental DC CB in a test laboratory in order to investigate the performance limits of the major internal sub-components of HVDC CBs. The lessons learned from experimental investigation are used to justify the test requirements specified for HVDC CBs.

1.5. Structure and Organization of the book

The remainder of this book is organized as follows. With the exception of Chapter 8, each chapter begins with brief review of state-of-the-art knowledge.

Chapter 2 provides theoretical background of DC short-circuit current interruption. The fundamental differences between AC and DC short-circuit current interruptions are highlighted.

---

[4]The works in this book are part of the EU funded project called PROMOTioN – PROgress on Meshed HVDC Offshore Transmission Networks [PRO]

Chapter 3 discusses various HVDC CB concepts and recent developments with a focus on the concepts which have made to prototype realizations. The actual status of development and the highest performances achieved for different technologies of HVDC CBs as reported in the literature are discussed in detail.

Chapter 4 presents system simulations studies of a conceptual MTDC grid. The temporal development, the main contributors and the critical factors determining the rate-of-rise and magnitude of fault currents are discussed in detail.

Chapter 5 The fault current interruption process by various models of HVDC CBs, and the critical stages and stresses on various components are identified in this chapter. The requirements of test circuits are also discussed.

Chapter 6 reviews various test methods and test circuits employed for testing HVDC CBs. The practical challenges associated with realization of different test circuits when evaluating with respect to rated performance demonstration is presented in this chapter.

Chapter 7 presents the novel test method proposed in this book. The mathematical background of the test method and an alternative test procedure in case of some practical limitations are discussed.

Chapter 8 presents the complete test circuit designed based on the proposed method, which is capable of providing all critical stresses to the HVDC CBs. The actual implementation of the designed test circuit along with the practical challenges and mitigation methods are presented with practical demonstrations in a test laboratory. The test results of several prototypes the HVDC CBs tested using the designed test circuit are presented.

Chapter 9 experimentally investigates the interaction of the internal components of HVDC CBs and the stresses on these components based on the test results of an experimental DC CBs set up in a test laboratory. The performances of various designs of vacuum interrupters while interrupting DC short-circuit current is studied. Also, the design and performance of the metal oxide surge arrester (MOSA) for HVDC CB application is studied.

Chapter 10 summarizes the conclusions based on the findings of the **book** and provides recommendations for future studies.

2. Fundamentals of DC Current Interruption

2.1. Introduction

In this chapter, the fundamentals of DC current interruption and the main differences between AC and DC current interruptions are discussed in detail. A mathematical analysis of DC current interruption at different stages of the process is provided. The analysis presented in this chapter provides fundamental background for the discussion in Chapter 3.

The requirements of DC current interruption and a few DC CB concepts have been extensively investigated in the 1970 - 1990. The fundamentals of DC current interruption remain the same, but the requirements on HVDC CBs have changed due to the progress in the HVDC transmission technology.

2.2. DC Current Interruption

The critical aspect that makes DC current interruption challenging is the fact that DC current remains constant in magnitude and direction – the DC current does not cross zero level. This means there is always magnetic energy stored in the system inductance – $\frac{1}{2}LI^2$, where L is the equivalent inductance of the system and I is the current flowing, see Figure 2.1. Under a short-circuit condition this energy can be very large as the short-circuit current increases. When an interrupter attempts to stop a DC current, the system tries to get rid of the magnetic energy in its inductance. Initially, this energy is transformed into electric energy by charging any capacitive element (C) that appears in parallel to the interrupter as shown in the figure. The ideal (lossless) relationship is as follows,

$$\frac{1}{2}LI^2 \Leftrightarrow \frac{1}{2}CU_{\mathrm{CB}}^2 \qquad (2.1)$$

Equation (2.1) shows that when an interrupter stops a DC current, a capacitor (stray or including a lumped element) is charged to a voltage (U_{CB}) that is proportional to the square root of the magnetic energy in the system. It is charged in such a way that its voltage is in opposite polarity to the source supplying the current – countering the source. It is the creation of this counter voltage that is the main principle in any DC current interruption.

In addition, the smaller the capacitor, the higher the charging voltage that appears across the interrupter. In reality the interrupter can withstand only limited voltage stress within a certain time of operation. Thus, if not somehow limited, the capacitor will be charged to an extremely high-voltage, which may exceed the dielectric insulation strength of the interrupter and cause damage to the system components[1]. A possible solution to reduce the charging voltage is by increasing the capacitance of the capacitor appearing in parallel to the interrupter. In fact, this becomes impractical as the energy in the system increases. A practical solution is to use a voltage limiting component, for example metal oxide surge arrester (MOSA), in parallel with the interrupter. In the latter case the excess magnetic energy in the system is transformed to thermal energy during the current interruption process.

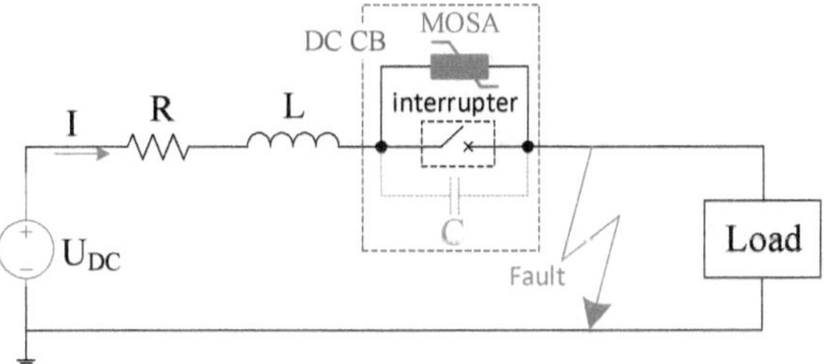

Figure 2.1.: A simplified DC circuit during a fault

An important conclusion here is that to achieve DC current interruption the counter voltage, henceforth termed as the transient interruption voltage (TIV), is essential and must be generated, limited to a safe level and maintained across the terminals of the interrupter. Nevertheless, to produce and maintain the TIV, the interrupter must be able to stop (locally) the DC current in the first place in order to commutate the current into the capacitor and MOSA. For this, a local current zero must be achieved in the interrupter.

---

[1]Even if the interrupter could sustain infinitely high voltage across its terminals, such interruption results in a very high voltage due to $L\frac{di}{dt}$ that results in the system.

Moreover, to suppress the system current, the TIV must be sufficiently higher than the source voltage, i.e., U_{DC} in the Figure 2.1. This is mathematically illustrated in Section 2.4. At low voltage, DC current interruption and suppression is achieved by an arc voltage that makes sufficient TIV (which is significantly in excess of the system voltage). In this case, the interrupter has to dissipate the system magnetic energy in the form of arc thermal energy. In order to enhance the arc voltage, intensive arc cooling mechanisms such as arc elongation and arc splitting have been used. However, the same approach cannot be applied at high voltage since the arc voltage that is produced in such a way cannot exceed a few kilovolts.

In general, a DC CB must perform the following fundamental functions to achieve DC current interruption:

1. Local current interruption

2. TIV generation

3. Energy absorption

It was quite clear from the early stages of HVDC fault current interruption studies that these functions should be separated in a DC CB [GL72]. There is no way that a single component can achieve these functions simultaneously.

2.3. AC vs DC Current Interruption

DC current interruption is fundamentally different from AC current interruption. Figure 2.2 shows current (black) and voltage (blue and red) waveforms during AC and DC current interruptions. The current in an AC system is sinusoidal, crossing the zero level at a rate of twice the power frequency. At each current zero the system inherently de-energizes itself i.e., there is no magnetic energy stored in the system inductance.

Figure 2.2a shows typical current and voltage waveforms during DC fault current interruption. At t_0 a short-circuit occurs in the system. This is followed by rapid rise of fault current, the rate-of-rise of which is limited by the system inductance, until steady state DC current is reached. Unlike in an AC fault current there are no naturally recurring current zero crossings in a DC fault current. The protection relay detects the fault and trips a DC CB at t_1. The DC CB performs its internal operation (involving local current interruption) and starts to generate the TIV at t_2. The internal actions of the DC CB, which precede the generation of the TIV, vary depending on the technology of DC CB. This is explained in Chapter 3. Then, the generated TIV is limited and maintained until the system short-circuit current is suppressed at t_3. The magnitude of the TIV is a design parameter of the DC CB, which must exceed the system voltage for which the breaker is intended. While maintaining the TIV during the interval between t_2 and t_3, the DC CB absorbs the magnetic energy from the system. Furthermore, the faster the DC CB produces the TIV following a trip command, the lower the peak short-circuit current in the system. Thus, a DC CB has a short-circuit current limiting feature. After current suppression at t_3, the DC CB is subjected to the DC recovery voltage from a system.

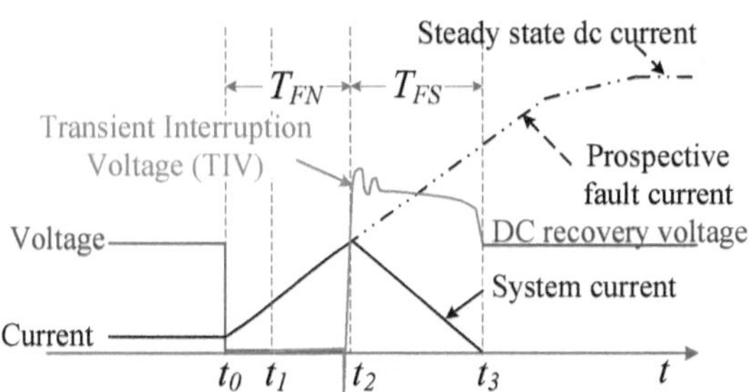

(a) DC current interruption waveform

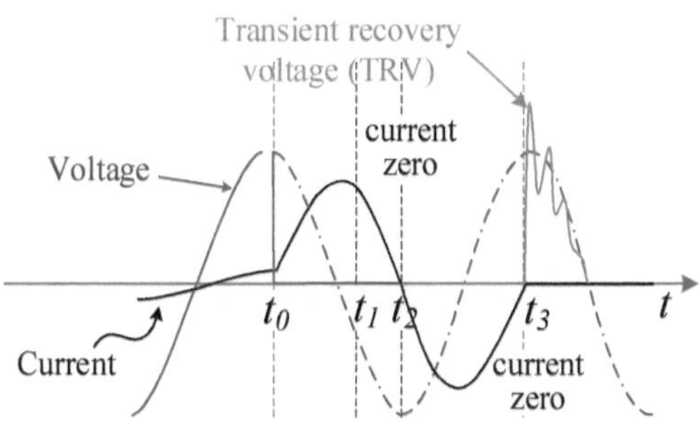

(b) AC current interruption waveform

Figure 2.2.: AC vs DC current interruption principles - waveform comparison

On the other hand, if no fast action is taken, the steady state DC short-circuit current, the magnitude of which is determined (on the DC side) only by inherent resistance in the circuit, continuously flows as shown by the black dash-dotted trace in Figure 2.2a. This can subsequently damage the system components which are predominantly PE semiconductors with a limited safe operation area (SOA[2]). In reality, the PE components in HVDC systems are protected well before the steady state DC short-circuit current is reached – for example by system control actions. The latter is explained in Chapter 4. However, to ensure continuous operation of these components (a converter as whole), a fast acting DC CB is indispensable so that the impact of a fault on the remaining healthy DC and AC systems is limited.

Generic current and voltage waveforms during AC fault current interruption are depicted in Figure 2.2b. When a short-circuit occurs at t_0, the current starts to rise with its peak value determined by the system voltage and circuit impedance up to the fault location. After a short while the protection relays detect the fault and send trip commands to the appropriate CBs. The AC CB operates its mechanism and opens its contacts at t_1 in Figure 2.2b. After contact separation, the system current flows through an arc established between the parting contacts. Current zero is reached naturally at t_2, where the arc is momentarily extinguished. This provides an opportunity for a CB to stop the continuation of the current flow by ensuring the arc between the contacts remains extinguished from that moment onward. For an AC CB to achieve current interruption at the first (natural) current zero (t_2), the contacts must have reached sufficient distance or fulfill a minimum arcing duration to withstand the transient recovery voltage (TRV). The latter is necessary in SF_6 based CB, in which sufficient SF_6 pressure must be built to extinguish the arc. If the arcing duration is too short or the contact gap is too small, the CB fails to clear and the arc re-ignites to re-establish current flow until the next current zero crossing which is at t_3 in Figure 2.2b. At t_3 the CB could withstand the TRV and subsequently the system voltage.

Table 2.1 summarizes the crucial features that distinguish a DC CB from an AC CB.

2.4. Mathematical Model of DC Current Interruption

Figure 2.1 shows a simplified DC system where an AC to DC converter is represented by an ideal DC source. The part shown as a load can be another DC to AC converter on the receiving end and is not relevant for the analysis in this section since it does not affect the operation of a DC CB after a short-circuit. For simplicity, it is assumed that a bolted short-circuit fault occurs near a converter terminal. Otherwise the effects of traveling

---

[2]The voltage and current conditions over which the device can be expected to operate without damage

Table 2.1.: Summary of comparison of AC CB vs DC CB

DC CB	AC CB
Active – limits the peak value of the short-circuit current	Passive – does not limit the peak value of short-circuit current
Produces, limits and maintains a TIV and imposes it on the system	The system determines the TRV and imposes it on the CB
Absorbs magnetic energy stored in the system inductance	Does not need to absorb the magnetic energy stored in the system inductance
Requires active power (Megawatts) source for testing – (traditional) synthetic testing method cannot be applied	Requires reactive power (Megavars) source for testing – synthetic testing method can be applied

waves need to be taken into account. Applying Kirchhoff's voltage law (KVL) to the simplified circuit in Figure 2.1,

$$U_{DC} - Ri(t) - L\frac{\mathrm{d}i(t)}{\mathrm{d}t} - U_{CB} = 0 \tag{2.2}$$

where, $i(t)$ is the fault current, U_{DC} is converter terminal voltage, R is inherent resistance in the circuit up to the fault location and U_{CB} is the voltage across a DC CB. Rearranging Equation (2.2), the rate-of-change of the short-circuit current can be described as follows,

$$\frac{\mathrm{d}i(t)}{\mathrm{d}t} = \frac{1}{L}(U_{DC} - Ri(t) - U_{CB}) \tag{2.3}$$

For mathematical convenience, the analysis is split into two consecutive time periods; fault current neutralization period ($T_{FN} = t_0 \rightarrow t_2$) and fault current suppression (energy absorption) period ($T_{FS} = t_2 \rightarrow t_3$) as discussed in [BS17], see Figure 2.2a. Actually a DC CB might have started to build the TIV a little earlier than at t_2. However, the period in which the voltage across the breaker rises to the desired level of TIV is assumed to be negligible for simplicity.

2.4.1. Fault current neutralization period – T_{FN}

This is the period from the fault inception at t_0 until the moment the TIV generated by a DC CB equals the system voltage; for example, at t_2 in Figure 2.2a. In this period, a DC CB does not influence the fault current since it is not inserting a counter voltage yet except a negligible arc voltage in some cases or a small voltage drop in some intermediate branches of some DC CBs. Referring to Equation (2.3), when the voltage across a CB is neglected i.e. $U_{CB} \approx 0$, the rate-of-change of current remains positive until $U_{DC} = Ri(t)$. Therefore, during this period, the fault current rises with its rate-of-rise limited mainly by the equivalent inductance of the system represented by L. Hence, neglecting the voltage across a CB and assuming initially steady state load current I_0 flowing before the short-circuit occurs, the solution of Equation (2.3) yields the prospective fault current described as,

$$i(t) = \frac{U_{DC}}{R}(1 - \exp(-t/\tau)) + I_0 \exp(-t/\tau) \tag{2.4}$$

where $\tau = \frac{L}{R}$ is the time constant of the system. From Equation (2.4), it can be seen that the prospective fault current increases to a steady state value of $\frac{U_{DC}}{R}$, if not quickly interrupted by a DC CB. Hence, on the DC side[3], the magnitude of the prospective fault current in DC systems is limited only by the intrinsic resistance R of the system.

2.4.2. Fault current suppression period – T_{FS}

This is an interval from the moment that the TIV equals the system voltage (at t_2) until the system current ceases to flow or falls below a residual value (at t_3), see Figure 2.2a. Assuming the U_{CB} is constant in this period and I_p is the value of fault current at the end of fault neutralization period ($I_p = i(T_{FN})$ in Equation (2.4)), the sign of rate-of-change of the short-circuit current (di/dt in Equation (2.3)) can be changed only if $U_{CB} > U_{DC}$. In other words, for a DC CB to achieve current suppression, it must produce a TIV higher than the system voltage. With this assumption, Equation (2.3) has the following solution [BS17].

$$i(t) = \frac{U_{DC} - U_{CB}}{R}\left[1 - \exp(-t/\tau)\right] + I_p \exp(-t/\tau) \tag{2.5}$$

Moreover, from Equation (2.5), the fault current suppression time (T_{FS}) can be determined by setting $i(t)$ to zero. In reality there could be a small leakage current after current suppression and this is neglected for simplicity. Accordingly, the fault current suppression time T_{FS} can be determined from the following expression [BS17, GDC11, KMMS68].

$$T_{FS} = \tau \ln\left[1 + \frac{R}{U_{CB} - U_{DC}}I_p\right] \tag{2.6}$$

Using Maclaurin series expansion of the natural logarithm, Equation (2.6) can be simplified as follows,

---

[3]if the DC source is an AC-to-DC converter, the fault current is limited by AC side impedance as well.

$$T_{\mathrm{FS}} = \tau \left[\frac{RI_{\mathrm{p}}}{U_{\mathrm{CB}} - U_{\mathrm{DC}}} \left(1 - \frac{RI_{\mathrm{p}}}{2(U_{\mathrm{CB}} - U_{\mathrm{DC}})} + \frac{1}{3} \left(\frac{RI_{\mathrm{p}}}{U_{\mathrm{CB}} - U_{\mathrm{DC}}} \right)^2 - \cdots \right) \right] \approx \frac{LI_{\mathrm{p}}}{U_{\mathrm{CB}} - U_{\mathrm{DC}}}$$

$$(2.7)$$

The approximation on the right hand side is obtained assuming an ideal situation where the total resistance in the circuit is neglected. This assumption makes the terms in parentheses of the middle equation converge to unity. Thus, Equation (2.7) shows that the fault current suppression time is directly proportional to the system inductance L, and the value of interruption current I_{p}, and inversely proportional to the voltage difference between the TIV and the system voltage.

Combining Equations (2.4) and (2.6), the fault current suppression time T_{FS} can be expressed as a function of system and DC CB parameters as follows:

$$T_{\mathrm{FS}} = \tau \ln \left[1 + \frac{U_{\mathrm{DC}}}{U_{\mathrm{CB}} - U_{\mathrm{DC}}} \left[(1 - \exp(-T_{\mathrm{FN}}/\tau)) + I_o \exp(-T_{\mathrm{FN}}/\tau) \right] \right] \qquad (2.8)$$

From Equations (2.6) and (2.8), it can be seen that the time required by a DC CB to suppress current I_{p} to zero depends on system as well as DC CB parameters. For example,

$$\lim_{U_{\mathrm{CB}} \to \infty} T_{\mathrm{FS}} \to 0 \qquad (2.9)$$

Equation (2.9) shows that the higher the TIV of a DC CB, the shorter the fault current suppression duration. However, the TIV is practically limited by the system as well as the DC CB insulation coordination. In addition the larger the inductance L in the system, the longer the T_{FS}.

The energy absorbed from the system during the current suppression period can also be described mathematically. Multiplying both sides of the KVL Equation (2.2) by $i(t)\mathrm{d}t$ and integrating over a period of T_{FS}[4], the following expression is obtained,

$$\int_0^{T_{\mathrm{FS}}} U_{\mathrm{DC}} i(t)\mathrm{d}t - \int_0^{T_{\mathrm{FS}}} R i^2(t)\mathrm{d}t - \int_{I_{\mathrm{p}}}^0 L i(t)\mathrm{d}i(t) - \int_0^{T_{\mathrm{FS}}} U_{\mathrm{CB}} i(t)\mathrm{d}t = 0 \qquad (2.10)$$

After some mathematical operation and rearrangement, Equation (2.10) results in the energy balance equation during the energy dissipation phase of the current interruption process. Accordingly, the energy balance in the system is given by the following expression,

---

[4] $i(t)$ changes from I_{p} to zero (negligible value) within the same duration

$$\frac{1}{2}LI_\mathrm{p}^2 + \int_0^{T_\mathrm{FS}} U_\mathrm{DC}i(t)\mathrm{d}t = \int_0^{T_\mathrm{FS}} U_\mathrm{CB}i(t)\mathrm{d}t + \int_0^{T_\mathrm{FS}} Ri^2\mathrm{d}t \tag{2.11}$$

The first term on the left hand side of Equation (2.11) represents the magnetic energy stored in the system inductance prior to the start of the fault current suppression whereas the second term on the same side represents the electrical energy injected by a DC voltage source (the rest of the system) during the fault current suppression period. The two terms on the right hand side of Equation (2.11) represent the energy absorbed by the DC CB and the energy dissipated in the circuit resistance, respectively. The latter can be ignored compared to the former since the energy dissipated in the circuit resistance R over a short T_FS is negligible. Thus, Equation (2.11) implies that a DC CB absorbs not only the energy stored in the system inductance but also the energy contributed by a DC source (system) during fault current suppression. The latter depends on the magnitude of the DC system voltage as well as the duration of T_FS as shown in Equation (2.13).

Assuming the ideal situation as in Equation (2.7) and solving the right hand side of Equation (2.11), the total energy absorbed by a DC CB can be computed as follows,

$$\int_0^{T_\mathrm{FS}} U_\mathrm{CB}i(t)\mathrm{d}t = \frac{1}{2}LI_\mathrm{p}^2 \left(\frac{U_\mathrm{CB}}{U_\mathrm{CB} - U_\mathrm{DC}}\right) = \frac{1}{2}LI_\mathrm{p}^2 + \frac{1}{2}LI_\mathrm{p}^2 \left(\frac{U_\mathrm{DC}}{U_\mathrm{CB} - U_\mathrm{DC}}\right) \tag{2.12}$$

Equation (2.12) can be further simplified as follows,

$$\int_0^{T_\mathrm{FS}} U_\mathrm{CB}i(t)\mathrm{d}t = \frac{1}{2}LI_\mathrm{p}^2 + \frac{1}{2}U_\mathrm{DC}I_\mathrm{p}T_\mathrm{FS} \tag{2.13}$$

Equations (2.12) and (2.13) show that the portion of energy that is contributed by a source during current suppression is proportional to the source voltage magnitude and inversely proportional to the voltage difference between the TIV and the source voltage. The higher the TIV relative to U_DC, the smaller the energy contributed by a source. In practical HVDC transmission systems, the TIV is limited to 1.7 p.u. by the insulation level of the transmission equipment [YTI$^+$82]. The minimum TIV considering the minimum interruption current is suggested to be at least 1.3 p.u. This is considering a system temporary overvoltage as a result of, for example, load shedding. To manage the energy absorption requirement of a DC CB, the contribution of the DC source in Equation (2.11) can be controlled. For example, by actively reducing the magnitude of the DC source voltage during fault condition (fault suppression) [GL72]. This is normally a common procedure in LCC HVDC transmission systems where the converter voltage is reduced by increasing the thyristor firing angle. It is even possible to make the source absorb the magnetic energy in the circuit by changing the polarity of the DC source. This is similar to AC power sources where this energy is oscillating back and forth between the source and the rest of the circuit, for example, by converting a rectifier into an inverter [EPR, FHKW69]. However, this action affects the normal operation of the healthy part of the system in a multi-terminal environment. In fact the preferred operation mode is indeed a "constant voltage" mode similar to AC systems [GL72].

Assuming U_{CB} is 1.50 p.u. of the system voltage (1.5 U_{DC}), which is the currently recommended practice, the energy contribution from the source is approximately twice the energy stored in the system inductance i.e. neglecting resistive losses during fault current suppression.

Therefore, combining Equations (2.9) and (2.11), it can be seen that the magnitude of the TIV also has an impact on the energy absorbed by a DC CB.

$$\lim_{U_{\mathrm{CB}} \to \infty} \left(\int_0^{T_{\mathrm{FS}}} U_{\mathrm{CB}} i(t) \mathrm{d}t \right) = \frac{1}{2} L I_{\mathrm{p}}^2 \tag{2.14}$$

Equation (2.14) shows the minimum energy absorption requirement of a DC CB that cannot be influenced even if the DC source can be shut off during fault current suppression.

2.5. DC Current Interruption Concepts

The above fundamental discussion shows that, to achieve DC current interruption, a DC CB must produce a TIV of sufficient magnitude and be able to dissipate the magnetic energy in the system inductance. In addition, it ideally needs to produce the TIV as quickly as possible after receiving a trip command to prevent the fault current from further increasing.

Therefore, in order to achieve DC current interruption, HVDC CBs employ multiple parallel branches as shown in Figure 2.3, each branch having components that serve various purposes during the current interruption process. The main branches are:

1. **Continuous current branch (CCB)**: A branch for nominal/load current conduction and local current interruption (to commutate current to another parallel branch).

2. **Commutation/current injection branch**: Branch(es) for reverse current injection, temporary current conduction, commutation and/or interruption and TIV generation.

3. **Energy absorption branch (EAB)**: A branch in which the TIV is limited, and maintained to a desired level and system energy is absorbed.

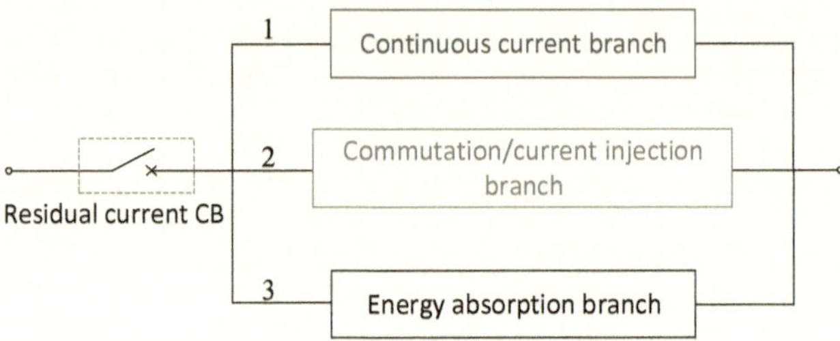

Figure 2.3.: Generic DC CB showing multiple current branches. ©[2018]IEEE.

In addition there is a residual current CB in series, (see Figure 2.3), which is an additional provision to isolate the HVDC CB from the system voltage after current suppression. This prevents leakage current through the EAB and/or through semiconductor switches (if they exist) when subjected to system voltage which, otherwise, could lead to thermal overload of these components.

The type and configuration of components in the first two branches, and especially the function and/or the number of current paths in the commutation branch, vary from one CB technology to another. Nevertheless, the main objective in any HVDC CB is to systematically commutate the DC current into an EAB under sufficient TIV condition.

In a nutshell an HVDC CB must fulfill the following basic conditions to successfully interrupt a DC fault current [Puc68, Fra11, BPS18b].

1. Achieve local current interruption/commutation in the CCB

2. Gain sufficient dielectric strength to withstand the voltage across its terminals during the current interruption process as well as the response from the network at later stage

3. Build up and sustain a TIV higher than the system voltage to drive the fault current to zero

4. Dissipate the magnetic energy stored in the circuit inductance and supplied by the DC source (system)

Hence, depending on the type and function of the main component(s) in the CCB and in the commutation branch, the state-of-the-art technologies of HVDC CBs can be classified into three major categories. These are,

1. **Solid-state (Power Electronic DC CB)** – PE switches interrupt system current, produce and sustain TIV.

2. **Mechanical HVDC CB** – mechanically opening contact system achieves local current interruption and sustains the subsequent TIV.

3. **Hybrid HVDC CB** – combined operation of mechanical and PE switches achieve local current interruption, produce and sustain the TIV.

Each of the above categories has associated pros and cons. For example, operating speed, maximum current interruption capability, cost, footprint/volume, complexity and operating losses as well as re-closing capability are the main distinguishing features among these technologies. The design and configuration of internal components, operation principles, status of development and finally a critical look at pros and cons of some of the state-of-the-art technologies of HVDC CBs are discussed in Chapter 3.

Even though a myriad of DC current interruption techniques has been proposed, only the HVDC CB concepts that have made way to high-voltage prototype developments and testing (at least at a proof of concept) are in the main focus of this **book**.

3. HVDC Circuit Breaker Technologies

3.1. Introduction

In order to identify the challenges associated with testing of various technologies of HVDC CBs, for example, any potential interactions with a test circuit, it is necessary to understand the behavior of these technologies during operation. Moreover, it is essential to define generic test requirements that address the critical aspects associated with the different technologies and also with high-voltage realizations. Armed with the fundamentals of DC current interruption discussed in Chapter 2, this chapter, therefore, reviews the actual DC current interruption techniques employed in the proposed HVDC CB technologies. The latest industrial concepts and prototype developments are reviewed along with their respective electrical designs and operation principles. The three categories of HVDC CBs described in Chapter 2 are discussed in detail.

3.2. Solid-State (Power Electronic) HVDC CBs

Static power electronic (PE) devices such as IGBTs, injection-enhanced gate transistors (IEGTs), bi-mode insulated gate transistors (BIGTs), insulated gate commutated thyristors (IGCTs) and gate turned-off thyristors (GTOs) are capable of changing state from conducting to insulating within a few to tens of microseconds using their gate controls. A generic model of a solid state DC CB is shown in Figure 3.1a. Essentially, there are two current branches, (1) a continuous current branch (CCB) and (2) an energy absorption branch (EAB) although there are additional snubber circuits for smooth switching of the semiconductors – the equivalent of which is shown by the green shadows in Figure 3.1b. Since there are no mechanically moving parts in these devices, they can stop DC current flow nearly instantly forcing the current to flow through the EAB made of MOSA. However, the PE switches have limited voltage and current rating, which means a large number must be used in series to withstand the TIV, and in parallel to cope with the large current interruption requirements [ST14].

Figure 3.1b depicts an example of a bidirectional solid state DC CB composed of a

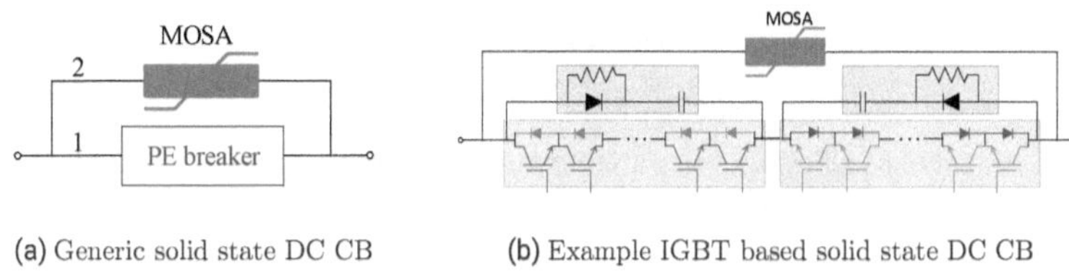

(a) Generic solid state DC CB   (b) Example IGBT based solid state DC CB

Figure 3.1.: Solid state (Power Electronic) HVDC CB

series connection of IGBTs in parallel with MOSA. Each IGBT has a snubber circuit for smoothing $\mathrm{d}u/\mathrm{d}t$ and $\mathrm{d}i/\mathrm{d}t$ stresses during turn OFF and ON, respectively.

The use of a large number of PE switches incurs two major technical challenges, the first of which is the increase in conduction loss during normal operation requiring active cooling [FGZ+16, WYC+18]. The second is the sophisticated and complex control (gate drive circuits) needed to continuously operate all the PE switches simultaneously. Furthermore, for a bi-directional current interruption, twice the number of PE switches are required since these switches have only unidirectional current blocking capability. These drawbacks limit the application of such a solution for use in an HVDC grid. The solid state DC CBs are more suited for low to medium voltage applications [ST14].

3.3. Mechanical (Local Current Zero Creating) HVDC CBs

The mechanical HVDC CBs consist of only mechanical CB(s) and passive components in the CCB, and auxiliary circuit needed for local current zero creation in the commutation branch. Hence, under normal operation, current flows through closed mechanical contacts of a CB. These types of HVDC CBs exhibit low conduction loss and are naturally bidirectional. A mechanical CB is responsible for local current interruption, which leads to the generation of TIV in the commutation branch. In fact for successful local current interruption, first, local current zero(s) must be created in the mechanical interrupter which is achieved using an auxiliary circuit in the commutation branch.

Based on current zero creation schemes, two categories of mechanical HVDC CBs exist. These are:

1. Passive oscillation current zero creation scheme

2. Active current injection current zero creation scheme – many variants exist

Especially in the active current injection category, several concepts have been proposed, developed and prototype tested, and a few are put in service. The most prominent developments are discussed below.

3.3.1. Passive oscillation – negative dynamic arc resistance

This is the earliest, the least complex, the most cost effective construction and the most exhaustively studied and prevalent DC CB concept to date [SKT+81, CVH+82, BMR+85, VCP+85, LSY+85, TAY+85, PMR+88, IHI+97, NNH+01, AH01, SKY14, WF14, JTWHW]. The electrical diagram of a passive oscillation DC CB is shown in Figure 3.2. It consists of a mechanical interrupter (AC CB) in the CCB and a series connected capacitor and inductor (L-C) in the commutation branch as shown in the figure. In addition, it has an EAB composed of a MOSA in the third parallel branch.

The interrupter is usually an AC gas CB, for example air-blast CB as in [BMR+85, VCP+85, PMR+88] and mostly SF_6 CB as in [LSY+85, IHI+97, NNH+01, AH01, JTWHW]. A gas CB is necessary because it exhibits negative dynamic arc (resistance) behavior which is essential during the current interruption process. The interaction of the L-C with the arc voltage of the interrupter excites a negatively damped resonant current that superimposes on to the arc current through the interrupter.

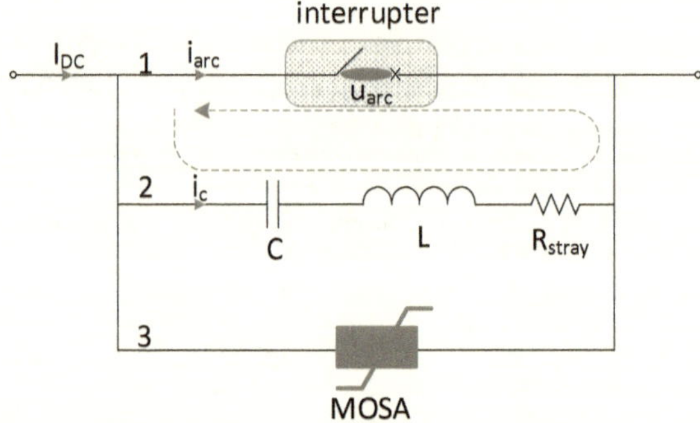

Figure 3.2.: Electrical diagram of passive oscillation mechanical HVDC CB

Referring to Figure 3.2, during the current interruption process, the characteristics of the arc voltage (u_arc) as a function of arc current (i_arc) follow a dynamic pattern dictated by the parameters of the main interrupter as well as the frequency of the oscillation branch. Thus, $\frac{\mathrm{d}u_\mathrm{arc}}{\mathrm{d}i_\mathrm{arc}}$ represents the arc's dynamic resistance. The arc current i_arc, when interrupting current I_DC, is given as in [PMR$^+$88],

$$i_\mathrm{arc} = I_\mathrm{DC} \left[1 + e^{-\left(\frac{\mathrm{d}u_\mathrm{arc}}{\mathrm{d}i_\mathrm{arc}} + R_\mathrm{stray} \right) \frac{t}{2L}} \sin \omega t \right] \tag{3.1}$$

where, $\omega = \frac{1}{\sqrt{LC}}$ is the angular frequency of the superimposed oscillating current, R_stray is the parasitic resistance of the circuit.

The second term on the right hand side of Equation (3.1) represents the oscillating part of the arc current which is superimposed. If the $\frac{\mathrm{d}u_\mathrm{arc}}{\mathrm{d}i_\mathrm{arc}} < 0$[1] and its magnitude is larger than the R_stray, then the amplitude of the self-excited oscillatory current increases exponentially. When the amplitude of the oscillating current equals (or exceeds) the I_DC, then current zero is achieved in the interrupter and hence, an opportunity for local current interruption is created. After local current interruption, the system current commutates, first to the commutation (L-C) branch and then, to the EAB after the capacitor in the commutation branch is charged to the TIV level determined by the MOSA.

Considering the recent advances in HVDC transmission technology, two main challenges prohibit this HVDC CB from use for a fault current interruption in MTDC grids.

1. **Very slow operation speed** – two main factors contribute to the slow operation. The first is the opening time[2] of a gas CB which could be as long as 20 ms [PMR$^+$88]. The second is the speed at which a current zero is achieved through the main interrupter once the arcing contacts separate. This depends on the magnitude of the interruption current as well as on the magnitude of the arc voltage and its characteristics.

2. **Limited maximum current interruption capability** – The maximum current interruption capability of passive oscillation HVDC CB is limited to a few kiloamperes [AH01], whereas the fault current levels in HVDC grids are expected to be significantly higher.

Hence, nowadays the passive oscillation HVDC CB is used mainly as a transfer switch rather than a CB. This includes neutral bus switch (NBS), earth return transfer switch (ERTS) and metallic return transfer switch (MRTS) – see Figure A.1. For these applications, recently, current transfer in excess of 6500 A has been achieved using double blast interrupter nozzle arrangement [BLRN, JTWHW].

---

[1] i.e., the arc exhibits negative dynamic resistance
[2] from the trip command until the contacts separate

3.3.2. Active current injection – direct discharge of pre-charged capacitor

The main drawbacks of the passive oscillation HVDC CB – slow speed of operation,
dependency on arc characteristics and limited maximum current interruption capability
– are addressed by using a pre-charged capacitor in the commutation branch as shown
in Figure 3.3a. The pre-charged capacitor is discharged to inject counter current which,
when superimposed onto the system current, creates local current zero in the interrupter.
In such a way the necessity for slow and bulky AC gas CBs has disappeared and instead,
fast and lightweight interrupters such as vacuum interrupters (VIs) can be used. Unlike in
the passive oscillation DC CB, the capacitor, in this case, remain charged and is isolated
from the circuit by a high-speed making device (see an example in Figure 3.3b) which,
when required, closes to discharge the capacitor.

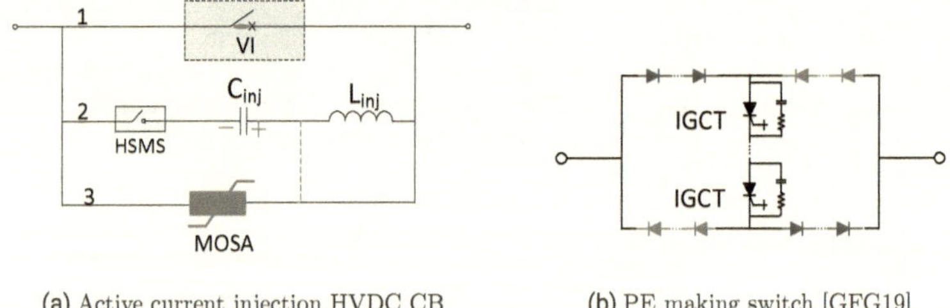

(a) Active current injection HVDC CB   (b) PE making switch [GFG19]

Figure 3.3.: Electrical diagram of active current injection HVDC CB – direct discharge of
pre-charged capacitor and a PE making switch

The concept is introduced for the first time in [GBK72, GL72] where a VI is used
as a main interrupter because of its superior performance at short contact separation.
The local current zero(s) can be created at any desired moment to enable local current
interruption by a VI, provided that its contacts have reached sufficient separation to,
subsequently, withstand the TIV. In fact the local current interruption does not stop
the fault current, rather it forces the current to commutate out of the interrupter into
a branch that generates a TIV. At this stage, the system current remains more or less
unaffected. However, from that instant on, the circuit current commutates first into the
parallel L-C branch which causes the voltage to rise until a level is reached that leads to
commutation to a MOSA in the EAB, see Figure 3.3a.

In addition, precise control and accurate timing of current injection is essential. In the
1980s, a high-speed mechanical making switch (HSMS) that can close within 3-4 ms was

inconceivable. Therefore, a simple closing switch like an ignitron or a triggered vacuum gap was used [GL72]. Triggered spark gaps are fast and precise, however, the continuous voltage stress from the pre-charged capacitor under normal operation condition poses a reliability issue. Recently, HSMSs operated by electromagnetic repulsion mechanisms are used instead. The latter suffer from pre-strike when closing under high-voltage, which causes current injection at an undesirable moment in the current interruption process. In order to mitigate this, high-speed, high-voltage vacuum making switches are used. Nevertheless, a number of HSMSs are needed in series connection to cope with the continuous stress from the pre-charge voltage.

In order to address the above issues PE making switches are proposed. PE making switches are the most accurate and reliable given their track record in other applications. However, a large number of PE devices[3] must be used in series to withstand the continuous voltage stress from the pre-charged capacitor. An example of a PE making switch, in this case, based on IGCTs, is shown in Figure 3.3b. A 500 kV active current injection HVDC CB using this PE making switch is realized for the Zhangbei HVDC grid in China [GFG19]. Further details about this breaker can be found in the Appendix A.4.

Recently, there is significant progress in the fast mechanical switching devices based on electromagnetic repulsion mechanisms [ABE+08, TKT+14, TSK+15, TIK+19, TIS+19, BHÁ+19]. As part of the PROMOTioN project, a prototype of an active current injection HVDC CB, rated at 160/200 kV, 16 kA, was tested in KEMA Labs. It has two series connected high-voltage VIs in the CCB. The counter current is injected from a pre-charged capacitor by closing two HSMSs (also vacuum) [TIS+19]. This breaker could demonstrate 16 kA current interruption within a breaker operation time of 7 ms while producing a TIV well over 300 kV. The test setup and results of this demonstration are discussed in Chapter 8 with further results in the Appendix A.6.2.

3.3.3. Active current injection – pulse transformer (coupled inductors)

During the early developments of an active current injection HVDC CB, the capacitor is charged to the line voltage from the DC bus. Some practical issues of such a design have been identified [YTI+82, STH+84]. The major issues are:

1. Spontaneous breakdown of gaps due to continuous high-voltage stress from the pre-charged capacitor over service lifetime

2. Long time storage of charges at high-voltage on the capacitor

---

[3]together with corresponding snubber circuits

3. Unavailability of HVDC CB during system energization because the capacitor
   cannot be sufficiently charged yet.

Especially, the triggered spark gaps become sensitive at high voltage and must be
carefully designed considering the basic insulation level (BIL) of the HVDC CB as
a whole. For example, for a 500 kV HVDC CB, the required BIL is 1550 kV. This
means, the triggered spark gaps must be designed to meet the BIL and also, it should
operate immediately upon reception of the trigger signals. To manufacture spark gaps
that can fulfill these two requirements at the same time is very challenging [STH+84].
Therefore, a design based on a pulse transformer (coupled inductor) was introduced in
[YTI+82, STH+84]. An electrical diagram of a recently revised pulse transformer based
design is shown in Figure 3.4 [LYX+19].

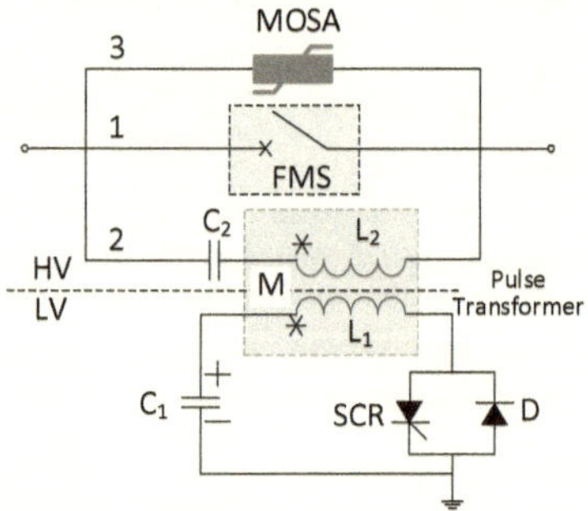

Figure 3.4.: Active injection HVDC CB based on coupled inductors (Pulse Transformer)
– adapted from [LYX+19]

In this design, a pre-charged capacitor is placed on the low voltage (LV) side of a pulse
transformer in order to mitigate the above issues. The counter current injection is applied
to the interrupter through a pulse transformer so that the HVDC CB can conveniently be
used in the system. In addition a re-closing feature can be supplemented relatively easily.
For example, two pre-charged parallel capacitors, each having an own closing device, can
be placed on the LV side of the pulse transformer for quick re-closing functionality. A
250 kV, 3500 A HVDC CB having the above features has been designed, manufactured
and tested in [YTI+82].

In the latest design a thyristor valve (silicon controlled rectifier (SCR)) with anti-parallel diode is used as a making switch instead of a triggered spark gap [LYX$^+$19]. Thus, the LV side has a SCR with anti-parallel diodes (D), a pre-charged capacitor (C_1) and LV inductance of the pulse transformer (L_1). Since thyristors are subjected to LV stress in this case, only a few series connected are needed. The HV side has capacitor C_2 (not pre-charged) and pulse transformer inductance (L_2). M is the mutual inductance of the pulse transformer. Thus, when the current in the CCB needs to be interrupted, the thyristors are turned ON with a continuous trigger mode. This ensures, together with the anti-parallel diodes, an oscillation creating multiple current zeros.

Sufficient number of current zero crossings is essential for the reliable operation of this concept. This requires proper design of the parameters on both LV and HV sides of the injection circuit shown in Figure 3.4. Especially, the pulse transformer must have a good coupling coefficient for efficient energy transfer from LV side to HV side during the current interruption process [LYX$^+$19]. In order to enhance the coupling coefficient as well as the insulation level, optimized oil immersed air core coupling reactors are proposed in [YTD$^+$19]. In fact, sufficient energy must be stored on the LV capacitor to create the necessary injection current peak on the HV side.

A 160 kV, 9 kA prototype based on this concept is recently developed, tested and put in service in the Nana'o radial HVDC pilot project in China [WWL$^+$18]. In this case four series connected fast VIs, each rated at 40 kV having own R-C//R grading elements, are used as the main interrupter in the CCB.

3.3.4. Active current injection – voltage source assisted oscillation

This is similar to the passive oscillation DC CB described in Section 3.3.1 in that an oscillating current with increasing amplitude is superimposed on the current through the main interrupter by an auxiliary L-C circuit. Similar to the passive oscillation DC CB, the resonant capacitor (C_R) remains uncharged during normal operation. However, the main difference is that the L-C resonant circuit, in this case, has a voltage source in series [Bon]. Thus, instead of self-excited oscillation based on the negative dynamic arc resistance of the main interrupter, the excitation of the oscillating current is driven by a voltage source in the commutation branch. The power frequency of the voltage source must be tuned to the resonant frequency of the L-C circuit. This approach alleviates many of the main challenges of the passive resonant DC CB among which are:

1. The maximum current interruption capability is not limited to a few kiloamperes since an oscillating current of large amplitude can be excited to create current zero.

2. Since the excitation is driven by an external voltage source, there is no requirement

of the main interrupter exhibiting negative dynamic arc resistance behavior nor high arc voltage.

3. Resonant current with rapidly increasing amplitude can be achieved by proper design of the resonance frequency and source voltage magnitude. Thus, faster current zero creation (within a few cycles) is ensured.

Hence, light weight VIs operated by fast actuators can be used as the main interrupter instead of bulky gas CBs. An innovative design based on this concept is proposed in [ANM16, ANMN17] as shown in Figure 3.5. This design uses a full-bridge voltage source converter (VSC) with a low-to-medium voltage (pre-)charged[4] DC link capacitor (C_{DC}) to excite the oscillating current in the L-C circuit. This principle is called VSC assisted resonant current (VARC) DC CB. Compared to the resonant capacitor (C_R), C_{DC} is very large i.e., $C_R \ll C_{DC}$ for sufficient energy storage[5]. The resonant components (C_R and L_R) are dimensioned at a frequency of a few tens of kHz in order to achieve rapidly increasing oscillation current. The frequency of the resonant circuit determines the switching frequency of the IGBTs in the VSC.

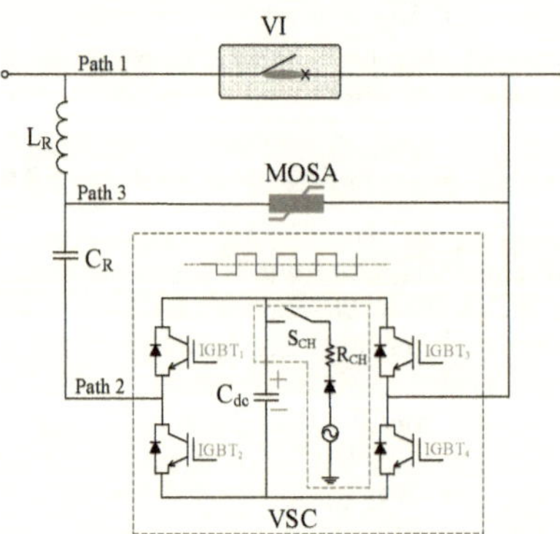

Figure 3.5.: Electrical diagram of VSC Assisted Resonant Current (VARC) DC CB

[4]The charging voltage of C_{DC} determines the number of series connected IGBTs in each arm of the VSC.
[5]Ideally, the VSC is a stiff voltage source for creation of the oscillating current with sufficiently increasing amplitude during the current interruption process.

The operation of the VARC HVDC CB during the current interruption process is as follows [ANM16, ANMN17, ABN$^+$18, ANM$^+$19]. Upon reception of a trip command, the VI opens its contacts which in turn is followed by arcing between the contacts. When the contacts of the VI reach sufficient separation, the VSC initiates the excitation of the resonance circuit to create current of increasing amplitude at high frequency. Referring to Figure 3.5, the VSC is operated in such a way that IGBT$_1$ and IGBT$_4$ are switched ON/OFF simultaneously whereas IGBT$_2$ and IGBT$_3$ are switched OFF/ON simultaneously at each current zero of the resonant current so that, at these switching instants, the polarity of C_{DC} voltage is in phase with C_R voltage – see Figure 3.6a and 3.6b. The resonant current is superimposed onto the main current through the VI. After a number of cycles, see Figure 3.6c, the resonant current grows to a value higher than (or equal to) the main current, thus creating a local current zero through the VI at which point local current interruption in the VI occurs. This forces the main current to commutate, first, to the commutation branch (path 2 in Figure 3.5) thus charging C_R until the clamping voltage of the MOSA is reached, subsequently leading to the system current suppression.

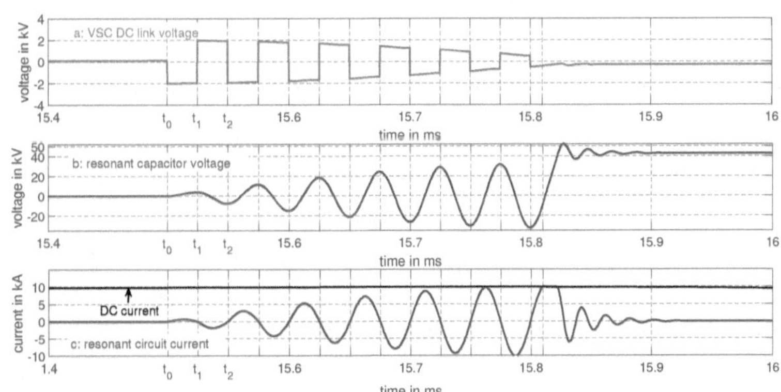

Figure 3.6.: Operation principle of VARC DC CB - resonant current excitation

An important feature of the VARC HVDC CB is that the current interruption is achieved when the resonant current is near its peak value despite the magnitude of the interruption current. Hence, the di/dt is not so severe, and since the C_R is almost discharged near the peak value of the excitation current, the ITIV is very low, as shown by the voltage across the VI trace in Figure 3.6b. Thus, both di/dt and the ITIV which are crucial success factors for current interruption adjust adaptively.

A 27 kV, 10 kA prototype unit of this concept has been developed and tested within the framework of the PROMOTioN project at KEMA Labs [ANM$^+$19]. A photo of the VARC DC CB prototype is shown in Figure A.15a. The prototype could achieve 10 kA current interruption within the breaker operation time of 3 ms while producing 40 kV TIV. Later, an 80 kV, 15 kA VARC HVDC CB consisting of three independent modules connected in series has been tested. In this case the breaker operation time of <2 ms has been achieved. The test setup and a few test results are discussed in Appendix A.6.3.

For extra high-voltage application (EHV) application, a large number of such units are connected in cascade. For example, 12 such modules are needed for 320 kV system voltage [LPM$^+$20].

3.3.5. Magnetic field controlled gas discharge tube based commutation branch

A different approach that provides current blockade without mechanical action is a gas discharge tube [HLRS76]. In fact, gas discharge tubes cannot conduct current continuously – hence can only form a commutation branch. Therefore, in the CCB, a parallel AC CB with fast opening speed having a few hundreds volts of arc voltage is used. An electrical diagram of a cross-field gas discharge tube based HVDC CB is shown in Figure 3.7. A DC CB based on such a concept has been developed and field tested in HVDC projects [HLK73, HLR$^+$76, GKV$^+$83].

During current interruption, the AC CB opens its contacts and at the same time the gas discharge tube is turned ON by applying cross-field (perpendicular to electric field) magnetic field. The arc voltage of the AC CB commutates current into the gas discharge tube which conducts current until the mechanical CB opens sufficiently. Once the mechanical gap gains sufficient dielectric strength, the magnetic field is removed which results in current quenching and a drastic increase in the arc voltage. This in turn commutates current from the gas discharge tube to a parallel capacitor which charges until the conduction voltage of the MOSA is reached.

At the time, it was found that the main limitation to the current interruption capability of this concept was the AC CB used in the CCB [CVH$^+$82]. The gas discharge tube could sustain a TIV of about 100 kV after current quenching although it is very sensitive to high du/dt. Further development of DC CB based on gas discharge tubes was halted for many challenging reasons [CVH$^+$82]. Nevertheless, recently the use of gas-discharge tubes has revived especially for a use in hybrid HVDC CBs [DBD$^+$, ABN]. In this case the gas discharge tube could be used in the commutation branch of hybrid HVDC CBs instead of a large number of PE switches, see Section 3.4.

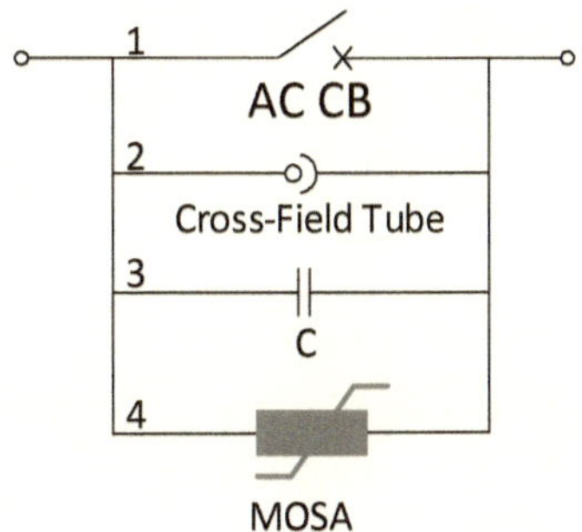

Figure 3.7.: Cross-Field Tube HVDC CB [HLR$^+$76]

3.4. Hybrid HVDC CB – Mechanical + Power Electronics

Hybrid HVDC CBs combine the desirable features of mechanical switches with those of PE switches. The desirable features of a mechanical switch are low conduction loss in a closed position and high-voltage withstand in an open position. However, the mechanical switches suffer from relatively long operation time[6]. The desirable features of a PE switch is high-speed operation which can be achieved within a few tens of microseconds. The PE switches, however, suffer from high conduction losses when carrying current and large numbers are required to withstand any practical high voltage in open (blocked) position.

Therefore, a hybrid HVDC CB is proposed in such a way that the CCB consists of a fast mechanical switch(es) in series with a (continuous) current commutation switch (CCS)[7] as shown in Figure 3.8. The CCS is needed for commutating the continuous/fast rising fault current into the commutation branch. The commutation branch is a PE breaker described in Section 3.2. This branch conducts the fault current until the (ultra-)fast mechanical switch(es) reaches sufficient dielectric withstand to protect the CCS against the subsequent overvoltage. Then, the PE breaker interrupts the fault current and this is

---

[6]both opening and closing operations

[7]In some cases the CCS is absent (hence indicated by dashed blue box in Figure 3.8) where current transfer to the commutation branch is achieved by arc voltage of the mechanical switch or a counter current injection feature. If the CCS is available, the mechanical switch does not interrupt current except a small residual and/or transient current ($<$5 A) caused by stray inductance during commutation.

followed by a steep voltage rise across the breaker, which is limited by the MOSA serving
the same purpose as in the mechanical type HVDC CBs.

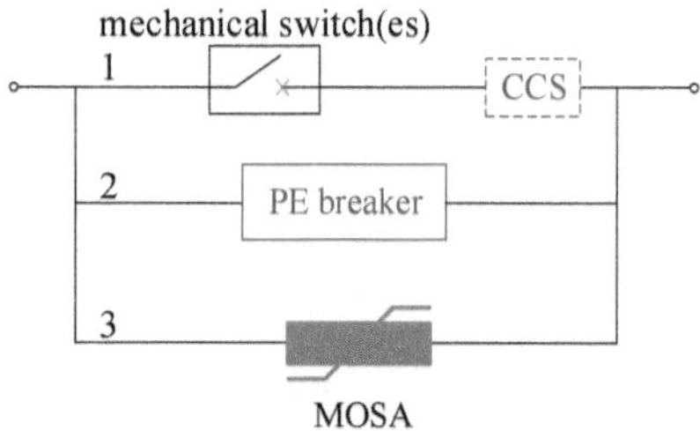

Figure 3.8.: Generic hybrid HVDC CB

The CCS is usually a matrix of several PE switches arranged in series and parallel (for
example, 3×3) in order to reduce the on state losses and increase reliability. Moreover,
these switches have a (water) cooling system since they conduct continuous current during
normal operation. However, no cooling system is required for the PE breaker in the
commutation branch since, under normal operation condition, no/negligible current flow
through this branch.

Depending on the type and arrangement of the PE switches in the CCS as well as in the
commutation branch, many variants of hybrid HVDC CB have been proposed, some are
prototype tested and a few have been commissioned in service [HJ11, GDPV14, ZWZ$^+$15].
Also, the fast mechanical switch could be SF_6 insulated disconnector(s) or a series
connection of VIs. Six different and most prominent topologies/realizations are discussed
below.

3.4.1. Hybrid HVDC CB – original topology

The electrical diagram of a hybrid HVDC CB – original topology is shown in Figure 3.9.
The CCB consists of an ultra-fast mechanical disconnector (UFD) and a CCS consisting
of a limited number of PE switches. In the commutation branch, it consists of a PE CB,
which is divided into breaker units (cells), sometimes known as the main breaker modules,
each of which can be operated independently. Each breaker unit consists of a cascade of
PE switches, which operate simultaneously and a parallel EAB (MOSA).

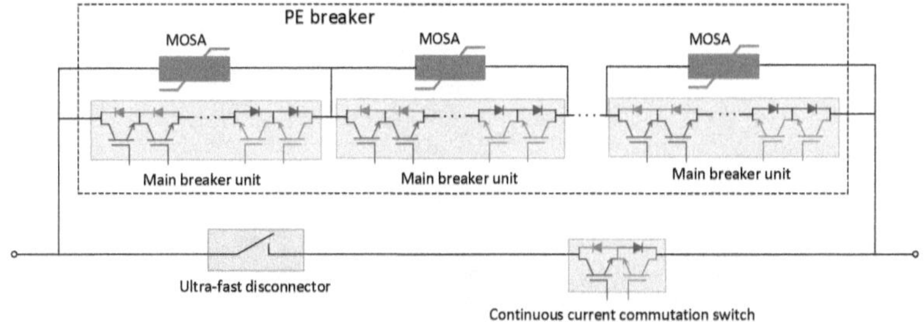

Figure 3.9.: Bidirectional hybrid HVDC CB – original topology

The fact that the breaker units have their own MOSA ensures equal voltage distribution across the units. Furthermore, this can be manipulated to produce a controlled TIV (staircase waveform) by successively switching these breaker units one after the other so that the overall du/dt stress on the UFD is reduced. In addition, the fact that the breaker units can be switched independently enables the possibility to be operated in a current limiting mode as discussed in [HJ11].

Recently a newly developed PE switch that integrates the anti-parallel freewheeling diode into the IGBT was proposed and used for HVDC CB application. The new development resulted in a single component, a reverse conducting IGBT, often referred to as Bi-mode Insulated Gate Transistor (BIGT)[8] [SRH+15]. The main claimed advantage of a BIGT is that it is compact and has higher current rating with up to double the interruption capability of the traditional IGBTs.

A prototype of a unidirectional hybrid HVDC CB built from BIGTs and rated at 350 kV (480 kV TIV), 20 kA, was recently tested at KEMA Labs. A breaker operation time of 3 ms is achieved. The laboratory test setup is shown in Figure A.8, and the electrical layout is shown in Figure A.11. Some of the test results are discussed in Chapter 8. Power supply at high potential to the gate units of the PE switches is challenging. In order to address this optically powered gate drives are used.

---

[8]The BIGT combines the functionality of a high-power IGBT and fast diode in the same silicon volume.

3.4.2. Hybrid HVDC CB – modified topology

One of the major drawbacks of hybrid HVDC CB discussed in subsection 3.4.1 is the fact that twice the number of PE switches required for unidirectional current interruption are needed for bidirectional current interruption capability. A few optimization efforts have been made to minimize the number of IGBTs required for bidirectional current interruption capability. An electrical diagram of one of the optimized designs is depicted in Figure 3.10. In this case, only a unidirectional PE breaker is required in the commutation branch while bidirectional current interruption capability is achieved by using a full-bridge diode rectifier as shown in the figure [WDB+].

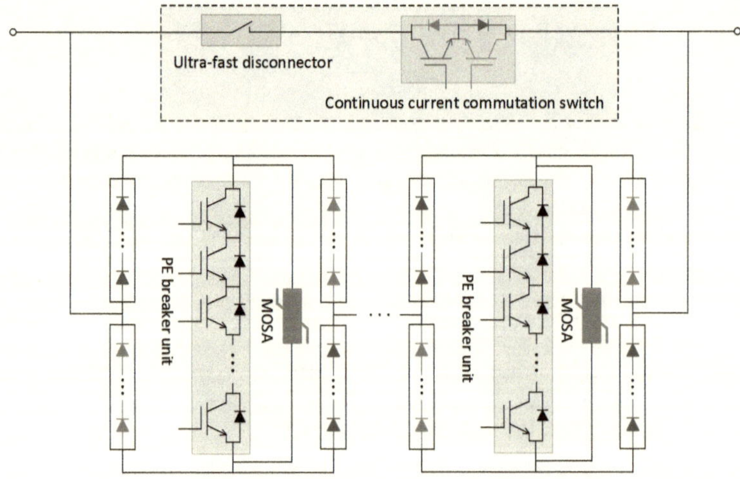

Figure 3.10.: Hybrid HVDC CB – Modified Topology [ZWZ+15]

The figure shows that the PE breaker (divided into several breaker units) is placed across the rectifier bridge. Hence, compared to the bi-directional hybrid HVDC CB shown in Figure 3.9, only half of the total IGBTs are needed in this case at the price of a full-bridge diode rectifier. During the current interruption process, the diode rectifier bridge ensures unidirectional current flow through the PE breaker despite the system current direction. The components of the CCB remain the same as for the original topology discussed above except, in this case, the UFD is built from a number of series connected VIs[9] [TZH$^+$]. When using a large number of series connected VIs, redundancy can be introduced in order to enhance reliability compared to a single high-voltage UFD. However, this comes at the expense of additional complexities such as voltage grading, synchronized operation and power supply to different units at high potential.

A 500 kV realization of the optimized hybrid HVDC CB (modified topology) is developed for application in the Zhangbei multi-terminal HVDC grid [TZH$^+$]. In this case the CCS is composed of 14 series × 6 parallel PressPack IGBTs with a commutation time less than 200 μs [TZH$^+$]. A UFD is composed of 10 series connected vacuum switches each of which are operated simultaneously by electromagnetic repulsion mechanisms. It is reported that within 2 ms these vacuum switches can achieve a total insulation level of 1000 kV with a voltage unbalance of less than 5%.

3.4.3. Hybrid HVDC CB – full-bridge IGBT based topology

A topology based on full-bridge sub-modules (SMs)[10], is depicted in Figure 3.11. The operation principle is similar to the other topologies described above. However, current interruption is achieved not by a cascade of single IGBTs, rather by a series connection of full-bridge SMs as shown in the figure. The CCB is composed of a matrix of full-bridge SMs (e.g., 2 × 3) connected in series with the ultra-fast mechanical switch(es). Likewise, the commutation branch consists of a large number of series connected full-bridge SMs designed to interrupt fault current and withstand the TIV. Finally, MOSA modules are connected across a few series connected SMs making one breaker unit. For example, one breaker unit is rated for 50 kV (80 kV TIV) and consists of 32 SMs [ZWZ$^+$15, JHR$^+$16, TWZ$^+$]. Multiple breaker units are connected in series to cope with the desired system voltage, for instance, four units are stacked for 200 kV system voltage.

Current in a full-bridge SM flows through two parallel paths (designated by blue (for forward) and red (for reverse) current directions) – each path being a series connection of an IGBT and a diode. Hence, the current breaking capability of such a design is twice that of the original topology (described in Section 3.4.1) built from the same type of IGBTs. In fact, such performance is achieved at a price of double the number of PE switches compared to the original topology.

Another important difference is that, unlike in the other topologies described above,

---

[9]The vacuum switches open under no current condition – thus arc-less operation is ensured since current commutation is performed by CCS before the contacts of the vacuum switches separate.

[10]A full-bridge SM consists four IGBTs, each having anti-parallel diodes, and a capacitor across the bridge.

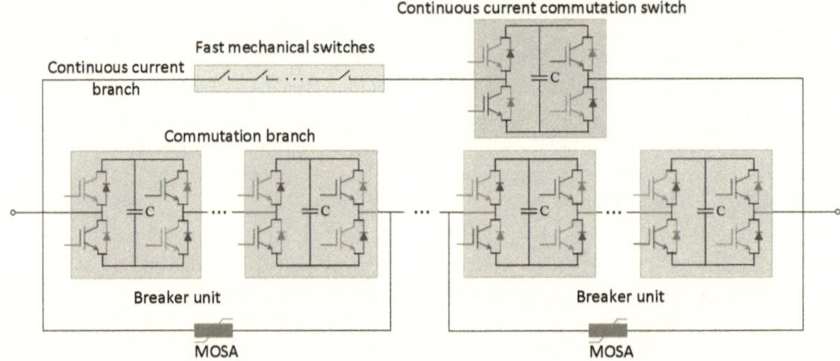

Figure 3.11.: Hybrid HVDC CB – Full-Bridge IGBT based Topology [ZWZ$^+$15, TWZ$^+$]

in this topology, the IGBTs do not require special snubber circuits since this function is provided by the SM capacitor. These capacitors are not normally charged like in a VSC, so a different form of isolated power supply to the gate drive units has to be developed, which is a challenge. The capacitors have associated discharging resistors connected in parallel (not shown in Figure 3.11).

The fast mechanical switch (FMS) is made of series connection of vacuum switches driven by electromagnetic repulsion mechanism (Thomson coil actuators) [ZWZ$^+$15]. Repulsive force produced by the discharge of pre-charged capacitors through a coil is used to open the vacuum switches. One of the main challenges of using such a large number of mechanical devices is synchronous separation of contacts as well as proper grading of the TIV during the current interruption process [YHW$^+$19].

The overall functionality test of a prototype consisting of one breaker unit in the commutation branch has been performed in a laboratory where 15 kA current interruption is reported within 3 ms [ZWZ$^+$15]. This topology has been put in service in the ±200 kV five terminal Zhoushan MTDC pilot project in China [THP$^+$15, TZH$^+$].

3.4.4. Hybrid HVDC CB – diode full-bridge topology

A hybrid HVDC CB described in subsection 3.4.3 is further optimized in a similar way as the original topology is optimized to the modified topology. In this case, the IGBT full-bridge SMs in the commutation branch are replaced by diode full-bridge SMs having (two parallel) IGBTs connected across the bridge as shown in Figure 3.12, [TZH$^+$].

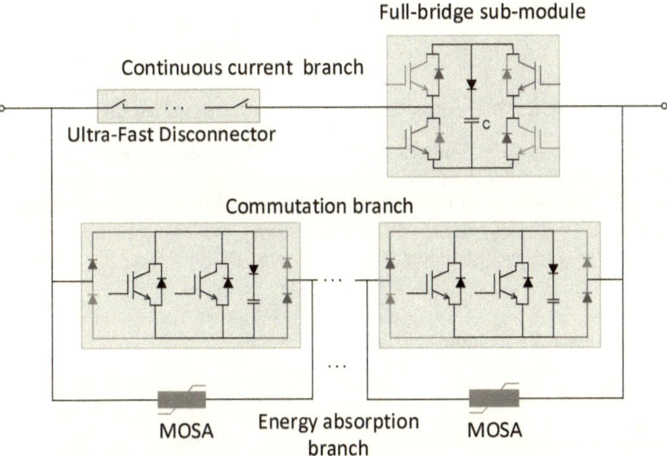

Figure 3.12.: Hybrid HVDC CB – Diode Full-Bridge based Topology [TZH$^+$]

The SM capacitor has a series connected diode across the bridge to prevent the IGBTs from stress during and after current interruption. Such arrangement results in the reduction of the required number of IGBTs by half while achieving the same current interruption performance as a IGBT full-bridge topology. The diode bridge redirects current in such a way that it flows through the IGBTs in the forward direction despite the system current direction. In such a way the size and cost of hybrid HVDC CB is further optimized. However, the CCB remains the same as described in subsection 3.4.3.

A 500 kV (800 kV TIV) hybrid HVDC CB based on this topology has been developed and functionally tested [TZH$^+$]. In this case a 2 × 5 matrix of IGBT full-bridge SMs is used in the CCB. However, in the commutation branch, diode full-bridge SMs are used as shown in the diagram of Figure 3.12. The commutation branch is composed of five breaker cells (modular units) each rated for 100 kV system voltage (160 kV TIV). The UFD is composed of 10 series connected VIs, each rated for 60 kV and 3300 A [TZH$^+$]. Functional tests of a complete prototype is reported in [TZH$^+$] stating that the hybrid HVDC CB could clear 26 kA current within 2.6 ms while generating a TIV of about 810 kV. Also, re-closing functionality has been tested where the breaker is re-closed 300 ms after the inception of the first short-circuit. Re-closing is achieved by turning the PE switches in the commutation branch ON. Re-closing of the CCB is performed only if the removal of a fault is confirmed. Otherwise, the re-opening is achieved by the commutation branch without involving the CCB. After re-closing, the breaker interrupted a current of 9 kA.

3.4.5. Hybrid HVDC CB – active current injection based topology

Further efforts to completely eliminate the operational losses associated with the CCS (and hence, the complexity associated with cooling system) in the CCB have been made. Two different designs, shown in Figure 3.13 and 3.14, are proposed to achieve this objective. Both concepts are based on the use of FMSs (CBs) with counter current injection features while still the system current is interrupted by cascade of PE switches in the commutation branch.

3.4.5.1. Negative voltage coupled (NVC)

This design, shown in Figure 3.13, is proposed in [WHL[+]18]. It completely removes the need of the PE based CCS. Current commutation to the commutation branch is achieved by the FMSs in the CCB. A pulse transformer based on coupled inductors[11] is placed in series with bi-directional PE breaker in the commutation branch as shown in Figure 3.13. The PE breaker consists of several series connected diode bridge SMs. Each SM has a parallel combination of R-C grading elements and MOVs connected across its terminals (not shown in the figure for clarity) [WHL[+]18, ZYZ[+]20, ZZXS20].

However, unlike the other hybrid HVDC CBs having PE switches in the CCS, the FMSs (VIs) in the NVC hybrid HVDC CB are subjected to arcing for a short duration (about 1.35 ms is reported in [ZZXS20]) – although the arc voltage is not sufficient for current commutation to the commutation branch. Thus, pulse current from a pre-charged capacitor is transferred via coupled inductors into the commutation branch to create a current zero through the opening/arcing FMSs. When current zero is created, local current interruption is achieved in the CCB. This forces current to transfer to the commutation branch.

Note that at current zero the FMSs do not see ITIV due to the residual charge on the pre-charged capacitor unlike in the mechanical active current injection HVDC CBs. Rather, only a voltage drop across the commutation branch is seen by the FMSs. After current commutation, the NVC hybrid HVDC CB behaves similar to hybrid HVDC CB

---

[11]Air core transformer having pre-charged capacitor on the primary side, a similar concept as described in subsection 3.3.3.

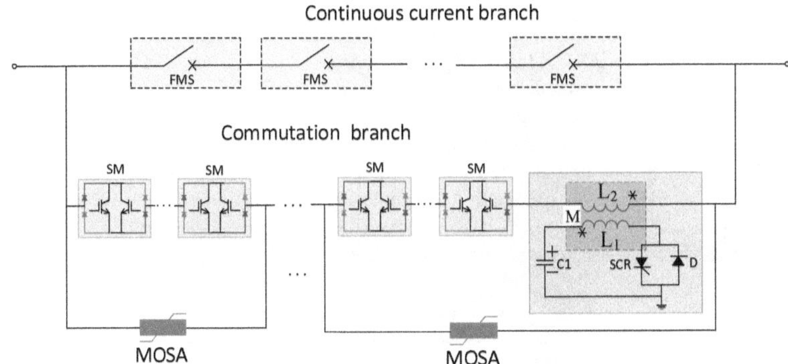

Figure 3.13.: Negative voltage coupled hybrid HVDC CB [WHL+18, ZZXS20]

described in subsection 3.4.4. Hence, once current is completely commutated to the commutation branch, the FMSs continue to open under no current and very low voltage condition (for another 1.35 ms) until it achieves sufficient dielectric strength to withstand the subsequent TIV when the PE breaker in the commutation branch block the current [ZYZ+20]. The counter current injection is precisely controlled by SCR[12] shown in the Figure 3.13.

One of the main challenges of the NVC hybrid HVDC CB is the efficiency in the coupling between the two inductors. In addition, because of the absence of the CCS, this design cannot have the proactive[13] current breaking feature.

A 500 kV prototype based on this concept is installed in a converter station of Zhangbei MTDC grid project [ZYZ+20].It consists of eight 100 kV FMSs and 320 bidirectional SMs in the commutation branch [ZZXS20]. The SMs in the commutation branch are divided into five breaker units each with common MOSA branch.

---

[12]The SCR of the coupling circuit is protected from over voltage by MOSAs - a possible source of which is induced voltage in L_1.

[13]an intermediate breaking process where current is commutated to the commutation branch before a trip command is sent by the protection system. The CCB is re-closed if the breaker does not receive a trip command

3.4.5.2. Parallel current commutation switches

In this design a PE commutation switch is placed in parallel with a mechanical CCS
(MCCS) [KHK+19, KOH+, DNA+17], see Figure 3.14. In the CCB there are two
mechanical switches, an ultra-fast disconnector and an MCCS as shown in Figure 3.14.
The former is required to withstand the full TIV while the latter is required to commutate
the continuous current under arcing condition. The MCCS is in parallel with an IEGT[14]
based full-bridge SM with pre-charged capacitor across the bridge. Under normal operation
condition, the IEGTs of full-bridge SM remain in the OFF-state; hence, no ON-state loss.
Compared to the NVC hybrid HVDC CB described above an added value of the latter
approach is that it has proactive current breaking feature.

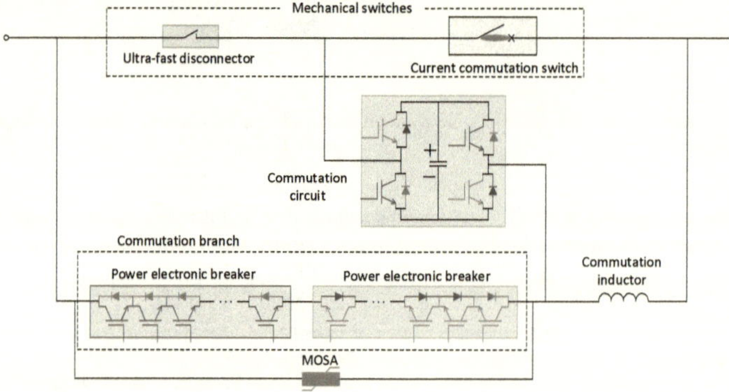

Figure 3.14.: Arc voltage based current commutation hybrid HVDC CB [KOH+]

If current interruption is required, a trip signal is sent to the breaker, after which the
breaker opens its mechanical switches. This is followed by arcing in both mechanical
switches. Then, the full-bridge SM, designated as commutation circuit (CC) in the
Figure 3.14, turns its IEGTs ON. Depending on the current direction either the bottom-
left and the top-right (colored blue), or the top-left and the bottom-right (colored red)
IEGTs are turned ON to commutate forward or reverse current, respectively. This
discharges the (pre-charged) SM capacitor of the CC through the commutation inductor

---

[14]The characteristics and performance of IEGTs in comparison with other PE switches is described in
    [CYZ+18].

shown in the figure to create a current zero in the MCCS. This results in the current commutation from the MCCS to the CC. Meanwhile the UFD switch is opening under arcing condition. Once current is fully commutated to the CC, the PE breaker in the commutation branch is turned ON and the IEGTs in the CC are turned OFF. This results in the charging of SM capacitor through the freewheeling diodes of the SM IEGTs. The voltage across the SM capacitor thus facilitates commutation of current from the arcing UFD to the PE breaker in the commutation branch. Following this step, the disconnector continues to open under no current/arcing condition[15].After dielectric recovery of the UFD, the PE breaker in the commutation branch turns its IEGTs OFF to produce the TIV after which current suppression starts in the MOSA.

One of the main differences between active current injection based hybrid HVDC CBs and mechanical HVDC CBs is that the counter current injection in the former case occurs well before the mechanical switches are fully open. This is because the TIV, in the former case, is delayed due to current conduction of the commutation branch unlike in the latter case where TIV generation follows immediately after local current interruption in the CCB. This means that the current injection circuits can be optimized since a high-peak counter current pulse may not be required to create a current zero at such an early stage of fault current development. In addition, components (capacitors and inductors) do not need to be designed for rated current and voltage.

3.4.6. Hybrid HVDC CB – thyristor based topology

This is a thyristor based design of a hybrid HVDC CB which consists of a CCB, several commutation branches – the time delaying branches, arming branch, and the EAB as shown in Figure 3.15, [GDPV14]. The figure depicts an electrical diagram a bidirectional thyristor based HVDC CB.

The main components in each branch are the following. The CCB is a low-impedance branch similar to that of the original topology hybrid HVDC CB. It consists of a UFD in series with IGBTs for commutating current into the first time-delaying branch. The time delaying branches consist of thyristor valves in series with capacitors, which, during the commutation process, charge sequentially to a voltage proportional to the dielectric strength of the parting contacts of the UFD[16]. In order to limit further increase in the voltage, the capacitors in the time delaying branches have surge arresters (SAs) in parallel, as shown in the figure. Besides, in these branches, relatively larger capacitors are used to limit the rate-of-rise of the voltage. When the UFD reaches sufficient contact separation

---

[15]Note that the system current does not need to commutate from the commutation inductor. The commutation inductor is required to limit the counter injection current peak value and frequency.

[16]The UFD starts opening by the time the current is fully commutated to the first time delaying branch.

(reported within 2 ms [GDPV14]), the thyristor valve in the arming branch (shown in path3) are latched ON and the capacitor C_3 starts charging. When the voltage across C_3 reaches the protective level of the main MOSA in the EAB (SA_3 in path4), the current commutates to this branch and the magnetic energy in the system is dissipated.

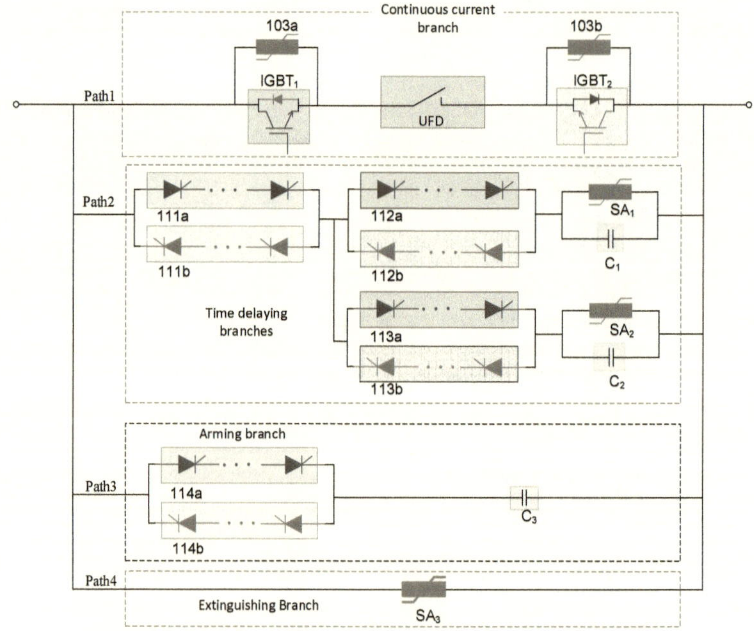

Figure 3.15.: Electrical diagram of thyristor based hybrid HVDC CB

The commutations in the several branches are important so that the breaker can handle the fault current until the TIV withstand level of the UFD builds to a required level in a controlled manner.

3.5. Summary and Conclusion

Based on the technology review of this chapter, it can be concluded that an HVDC CB is not just a mechanical contact system, rather a system of components arranged in multiple current branches, to which current is commutated in a controlled manner to achieve DC current interruption. For (extra-) high-voltage applications, HVDC CBs are designed in a modular approach, where several components are combined in series either per current branch(es) or as independent HVDC CB modules. This adds to the complexity and footprint of such equipment.

Two leading technologies; namely, the mechanical active current injection HVDC CB and the hybrid HVDC CB have become the preferred candidates for MTDC grids. The gap in performance between these technologies is narrowing, and in some cases similar performance requirements have been set and met. Several realizations of each category with up to 500 kV system voltage, <3 ms breaker operation times and 25 kA current interruption are developed, and some are applied in the recent pilot projects. Although full PE HVDC CBs are possible in theory, the conduction losses during normal operation prohibits the practical application at high voltage.

The mechanical active current injection HVDC CBs provide low loss during normal operation and are naturally bidirectional. However, long duration energy storage on capacitor banks at high potential (and hence, its isolation during normal operation), the need for HSMSs, requirement for additional provisions required for re-closing operation and prone to re-ignitions are some of the main challenges. Measures to address some of these challenges as in coupled inductor and VARC technologies have been proposed but with associated complexities and limitations.

The hybrid HVDC CBs have operational power loss and hence, cooling requirement of CCB. In addition, the need for large number of series connected PE switches in the commutation branch, complex snubber circuits and power supply to these switches at different potentials are some of the challenges. Some efforts have been made to optimize the number of the PE switches required in the commutation branch.

In all HVDC CBs a fast mechanical switchgear is present either to interrupt and/or isolate. This component determines the speed of operation which is a crucial parameter of the breaker. To achieve fast contact separation, electromagnetic pulse drive mechanisms are used instead of conventional spring, hydraulic or magnetic drives. To cope with high-voltage withstand in open position, several series connection is necessary. In this case synchronous operation (opening), equal voltage distribution and power supply at different potentials are the main challenges.

4. Analysis of Faults in Multi-Terminal HVDC Grids

This chapter focuses on fault conditions in MTDC grids. The discussion of the chapter is based on the results published in [BPS18b, BSPN17]. The chapter is organized in different sections. First, brief background of the chapter is presented in Section 4.1. Then, MTDC simulation systems in the literature are briefly reviewed in Section 4.2. The benchmark system used for the simulation study of this **book** is described along with the underlying assumptions in the same section. Then, a detailed discussion of HVDC fault conditions is presented in Section 4.3. The analysis of the converter's response to a DC fault and a fault current generic to half-bridge MMC VSC based MTDC grids are presented. The effect of various critical network parameters that have an impact on the fault current are discussed. Finally, the summary and conclusion based on the results of the study in the chapter are provided in Section 4.4.

4.1. Introduction

Since its inception, HVDC transmission has advanced through several technological steps – from mercury arc to thyristor based LCC followed by IGBT based VSC offering several new features and possibilities. The latter itself has evolved significantly – from a simple two-level converter to the state-of-the-art MMC. Each technology has different vulnerabilities and in particular, a different response to HVDC grid faults, which entails considerably varying degrees of requirements for HVDC CBs. This, coupled with a lack of operational experience, has made the definition of the scope and duty of HVDC CBs challenging.

In the absence of operational experience, the effective approach to determine the requirements of an HVDC CB is to perform system simulation studies. For this, availability of an adequate simulation platform and a sufficiently detailed system model with realistic system parameters is necessary. In fact the use of system simulation studies is not new. For example, simulation study of an actual system was performed in the 1980s to determine the specifications of HVDC CBs [STH$^+$84, P MR$^+$88]. Although the accuracy of simulations was limited back then, it was understood that the HVDC CB needs to clear not only the maximum fault current but also the load current of the system. This is due to the fact that the interruption of the load current is more severe especially with respect to mechanical interrupters. Moreover, energy absorption is identified as an essential part of an HVDC CB's duty and it was evident that the latter is affected by many factors including converter control and system configuration [YTI+82].

Considering the stringent requirements of the recently developed VSC HVDC transmission schemes, several manufacturers have proposed and developed prototypes of HVDC CBs as discussed in Chapter 3. In order to specify the requirements of an HVDC CB, first, it is essential to identify the major factors that determine the fault currents in a meshed MTDC network. Then, to investigate the worst-case stresses to which the HVDC CBs are subjected considering the actual operation conditions. The former is the focus of this chapter while the latter is discussed in Chapter 5.

4.2. Multi-Terminal HVDC System Studies

4.2.1. Review of benchmark grids

Due to lack of actual VSC MTDC grids in operation, so far research is limited to simulation studies using different conceptual HVDC networks. This research is mainly aiming:

- Load flow calculations – to determine optimal power flow control and dispatch [WA12, BWAF14, BCB14, AHWT16, ATW17]

- Fault detection and investigation of protection strategies [LH14, GLH15, SR16]

- Analysis of system transients – evaluation of rate-of-rise and magnitude of fault currents, fault current contributions, impact of system grounding schemes, etc., [BF13b, BF13a, TKT+14, LTBH14, BWAF14, BF14, CWM+18]

Driven by various purposes, a multitude of MTDC grid architectures along with varying number of converters and system parameters have been used as benchmark study grids. Different converter topologies (two-level, multi-level, etc.), converter configurations (symmetric/asymmetric monopole, bipole) and various grounding schemes (solid, impedance) have been investigated. In some cases different voltage levels with DC/DC converters are used [ATW17]. Different MTDC grid configurations (radial, ring, star and meshed) are considered. For example, various grid configurations for integration of large offshore wind-farms are evaluated and compared in terms of features such as flexibility, redundancy, line lengths and ratings, the need for communication as well as the number of HVDC CBs required in the networks [GBLE+11].

Also the level of detail and hence, the accuracy of simulation models vary significantly depending on the purpose of study. Hence, the results obtained in one study differ considerably from the other, and this makes it difficult to make proper comparisons and draw conclusions. For example, considering fault analysis, many different assumptions exist. In some studies initially blocked converters[1] are considered [BF13b] whereas in some others it is assumed that a converter does not block at all during faults [TKT+14]. In reality a converter blocks only after a fault is detected in the system. In a point-to-point system a converter blocks as soon as a fault is detected. This is desirable because, in any case, a fault leads to a total loss of power transmission on the faulted pole, whereas in VSC MTDC grids, in which selective protection is sought, the preferred option is to isolate the faulted section by HVDC CBs before any of the converters blocks. The latter results in the least disturbance to the system compared to blocking a converter during a fault condition and unblocking it after a fault clearance. This is illustrated via simulations in Section 4.3.3.

In most cases, a generic study showing the detailed phenomena during DC faults, independent of network topology, is missing. For example, fault current testing envelopes for HVDC CBs have been investigated for different converter topologies under various fault conditions in [CCB+16]. Several design considerations of HVDC CBs such as the energy absorption requirement, the minimum DC current limiting reactance required to prevent a converter from blocking, etc., have been studied in [DEW+16, CWM+18]. However, these studies mainly focus on fault currents supplied by a converter and do not consider the impact of the in-feed from multiple adjacent line/cable connections in an MTDC environment.

Simulation studies show that grid topology and converter configuration have a significant impact on the requirements of HVDC CBs [KPRB15, BWAF14]. An investigation of fault current contributions of various elements as well as the impact of different grounding schemes have been conducted [BF13a, BF14, LTBH14]. For example, a bipolar converter represented by a simple diode rectifier, assuming a blocked converter during a fault, is simulated. Moreover, a four-terminal meshed MTDC grid based on the same assumption is used for the investigation of the current interruption processes in various HVDC CB technologies [BF16]. However, the fact that all converters are represented as blocked converters during a pole-to-ground fault is not realistic since in practice only the converters on the faulted pole and only stations located close to the fault are affected. In addition, this assumption has insufficient detail of the events occurring during the transition from normal operation to fault condition because of the lack of a control system.

A four terminal meshed grid built from symmetric monopole converters is proposed as a

---

[1]the equivalent of which is simple diode rectifiers

benchmark network in [LH14, LAH$^+$15]. A similar network topology but at reduced power and with bipolar converter configuration is considered in [HT13]. Another four-terminal radial network based on symmetric monopole converters is used to study the impact of various DC current limiting reactors (L_{DC}) on the rate-of-rise of DC fault currents [TKT$^+$14].

Acknowledging the importance of having a unified benchmark system for harmonizing the research related to MTDC grids, Cigré WG B4-57 has developed an 11-terminal HVDC system in [Cig14]. The system is shown in Figure A.3 in the appendix. Detailed data and control parameters for steady state power flow have been provided.

4.2.2. Reduced MTDC benchmark grid – system description

The 11-terminal benchmark grid introduced by Cigré WG B4-57 is complete, however, still computationally intensive to perform transient simulation studies especially at smaller simulation time steps. Thus, for the purpose of studying faults in MTDC grids in this book, a hypothetical five-terminal meshed offshore HVDC grid shown in Figure 4.1 is used as a benchmark system instead. All converters are bipole converters operated at ±320 kV with earth return. In fact the system data for this network is adopted and modified from the DCS3 of the Cigré benchmark system shown in Figure A.3 [Cig14].

Three of the five bipolar converters, namely, C-C2, C-D1 and C-D2 are assumed to be located offshore representing offshore wind-farms with generating capacity of 600, 1200 and 900 MW respectively, and are interconnected via submarine cables in a radial arrangement. These converters are connected to wind-farms which are represented by a weak AC grid with short-circuit capacity (SCC) of 3.8 GVA[2] at the point of common coupling (PCC)[3]. The remaining converters, namely, C-A1 and C-B1 are located onshore and receive power from the offshore network via three DC cables as shown in Figure 4.1. The onshore converters are connected to a strong AC network with SCC 30 GVA. For simplicity, all AC sources including the wind-farms are modeled as a voltage source behind an equivalent impedance taking into account the short-circuit power.

$$SCC = \frac{U_{AC}^2}{X} \tag{4.1}$$

---

[2]Depending on the type of wind turbines, the actual fault response of a wind-farm is much more complex than this simplified assumption, but this is ignored for simplicity in this study. The wind turbines actually require an AC voltage to operate, but since type 4 wind turbines are all converter connected, it is reasonable to assume that their fault current contribution is no more than 2 p.u., thus, a weak AC grid.

[3]at AC side of the converter transformer

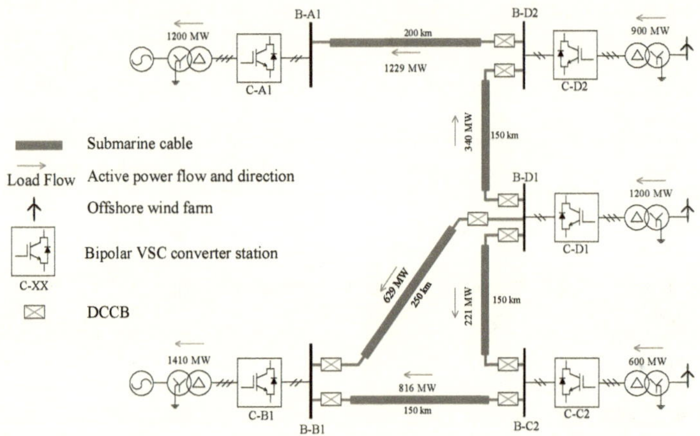

Figure 4.1.: Hypothetical five terminal bipolar HVDC grid.

Where U_{AC} is the line-to-line voltage of the generator and X is the impedance up to the PCC. Equation (4.1) is assuming negligible resistance in the circuit.

Nowadays, MMC VSC has become the preferred converter topology because of its advantages compared to other VSC converter topologies [SDM+14, MB15]. Several recently commissioned HVDC projects are based on MMC [Tee, PSD+, Irn20]. Hence, all converters of the benchmark grid are of MMC type with half-bridge SMs. The control modes of each converter shown in Figure 4.1 is as follows. Converter C-A1 controls DC voltage; thus, serves as slack bus. Converter C-B1 is set to control active power versus DC voltage (PV droop control). The offshore converters (C-C2, C-D1 and C-D2) operate as rectifier and control active power injected into the DC network. The system parameters of each converter are shown in Table 4.1.

A frequency dependent (phase) model of a cable implemented in PSCAD ([MGT99]) with parameters obtained from Cigré Technical Brochure 604 [Cig14] is used in the simulation. The arm reactors (L_{arm}) and the SM capacitors (C_{SM}) are dimensioned according to the equations provided in the Technical Brochure.

---

[4]makes 0.15 p.u.

[5]Nearest Level Control

Table 4.1.: Simulation system parameters

Parameter (unit)	Symbol	C-A1	C-B1	C-C2	C-D1	C-D2
Rated power per pole (MVA)	S	800	800	400	800	600
Active power set point (MW)	P	600	700	300	600	450
Arm reactor (mH)[4]	L_{arm}	29	29	58	29	38.5
Transformer leakage reactance (p.u.)	L_{tr}	0.18	0.18	0.18	0.18	0.18
Transformer series resistance (p.u.)	R_{tr}	0.6	0.6	0.6	0.6	0.6
Primary voltage (kV)	U_{p}	380	380	145	145	145
Secondary voltage (kV)	U_{s}	220	220	220	220	220
AC system frequency (Hz)	f	50	50	50	50	50
DC voltage (kV)	U_{DC}	±320	±320	±320	±320	±320
Number of SMs per arm	N_{SM}	160	160	160	160	160
SM capacitor (μF)	C_{SM}	12500	12500	6250	12500	9375
Modulation technique	-	NLC[5]	NLC	NLC	NLC	NLC
Converter losses (%)	-	0.3	0.3	0.3	0.3	0.3

In this study, only a pole-to-ground fault is considered. A pole-to-pole fault for symmetric monopole converter is investigated in the literature such as [TKT+14, CWM+18]. In fact there is significant resemblance between a pole-to-ground fault in a bipolar system and a pole-to-pole fault in a symmetric monopole system. The difference in terms of fault current and system voltage is attributed to the grounding type and impedance in the bipolar system. In addition, unless otherwise explicitly mentioned, a fault is applied on a cable between converters C-D1 and C-B1 at a distance of 10 km from C-D1. In order to reduce the rate-of-rise and hence, limit the magnitude of the fault current to remain within the breaking capability of the recently developed technologies of HVDC CBs, L_{DC} are placed in series with the CBs. The L_{DC} not only reduce the rate-of-rise of fault current through the CB but also delay the propagation of the impact of the fault (voltage collapse) within a grid. Also for fast, selective and reliable protection algorithms L_{DC} is crucial [WBMC16, LXY17, SR16, LTF17]. Thus, in this chapter, unless otherwise explicitly stated, an L_{DC} of 100 mH[6] is placed at the ends of each DC cable.

4.3. HVDC Fault Current Analysis

System startup is not part of this study. A fault is applied after the system has reached steady state operation. A description of hierarchical control and a mathematical descrip-

---

[6]Depending on the type of HVDC CB and protection strategy, different values may be used in reality. The presence and value of series reactors can also have an adverse effect on the controllability of a grid, in addition to the usual aspects such as additional cost, space, losses and maintenance.

tion of steady state operation of an MMC converter is discussed in detail the literature [SDM$^+$14, Cig14].

4.3.1. Fault conditions in MTDC grids

A short-circuit on a power line (cable or OHL) provokes traveling waves that propagate along the line in both directions of the fault location. The magnitude of the wave depends on the fault resistance [BF13a]. The voltage waves travel along the cable, while discharging the distributed capacitance along its way, towards the DC buses at either ends of the faulted cable, where converters and/or other links are connected. Considering a pole-to-ground fault on the positive pole in a bipolar HVDC grid interconnected with cables, the incident traveling wave first arrives at L_{DC} at the ends of the faulted cable, before propagating into the other elements connected to the DC bus. A part of this incident wave is transmitted through L_{DC} while a large part of it is reflected.

To illustrate this phenomenon, a simplified equivalent system of the two sides of L_{DC} is shown in Figure 4.2. Assuming the incident voltage wave as a (negative) step function

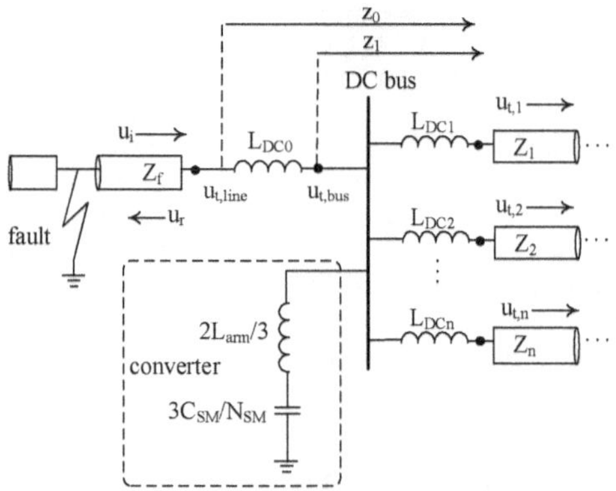

Figure 4.2.: Simplified system diagram demonstrating the impact of the current limiting reactors (L_{DC}) during a fault. ©[2018]IEEE.

for simplicity, the part of the wave that is transmitted through the L_{DC0} is given by the following mathematical equation (in the Laplace domain),

$$u_{t,line} = \frac{2z_0}{z_0 + Z_\mathrm{f}} u_\mathrm{i} = \frac{2(z_1 + sL_{\mathrm{DC0}})}{z_1 + sL_{\mathrm{DC0}} + Z_\mathrm{f}} u_\mathrm{i} \tag{4.2}$$

Where z_1 is the equivalent impedance of the system seen at the DC bus side, and $z_0 = z_1 + sL_{\mathrm{DC0}}$ is the equivalent impedance of the system seen at line side of the L_{DC0}, Z_f is the characteristic impedance of the faulted cable and u_i is the magnitude of incident wave (see Figure 4.2) and $s = j\omega$.

Considering the right-hand side of Equation (4.2) in reference with Figure 4.2, the first term in the numerator represents the portion of the incident wave going into the DC bus whereas the second term represents the portion of the transmitted voltage wave dropped across the L_{DC0}. Solving Equation (4.2), the transmitted voltage wave in the time domain,

$$u_{t,line} = u_\mathrm{i} \left(\frac{2z_1}{z_1 + Z_\mathrm{f}} \right) \left(1 + \frac{Z_\mathrm{f}}{z_1} \exp\left(-\frac{z_1 + Z_\mathrm{f}}{L_{\mathrm{DC0}}} t \right) \right) \tag{4.3}$$

From Equation (4.3) the part that is propagating into the system and thus causing the discharge of any capacitive elements including SM capacitors; and the part that appears across the L_{DC0}, respectively, are

$$u_{t,\mathrm{bus}} = u_\mathrm{i} \left(\frac{2z_1}{z_1 + Z_\mathrm{f}} \right) \left(1 - \exp\left(-\frac{z_1 + Z_\mathrm{f}}{L_{\mathrm{DC0}}} t \right) \right) \tag{4.4}$$

$$u_{L_{\mathrm{DC0}}} = 2u_\mathrm{i} \exp\left(-\frac{z_1 + Z_\mathrm{f}}{L_{\mathrm{DC0}}} t \right) \tag{4.5}$$

Furthermore, the portion of the voltage wave propagating into cable i connected to the DC bus is,

$$u_{t,\mathrm{i}} = \frac{Z_\mathrm{i}}{Z_\mathrm{i} + sL_{\mathrm{DCi}}} u_{t,\mathrm{bus}} = \left(\frac{Z_\mathrm{i}}{Z_\mathrm{i} + sL_{\mathrm{DCi}}} \right) \left(\frac{2(z_1 + sL_{\mathrm{DC0}})}{z_1 + sL_{\mathrm{DC0}} + Z_\mathrm{f}} \right) u_\mathrm{i} \tag{4.6}$$

Equations (4.2)-(4.5) show that the larger the inductance L_{DC}, the smaller the portion of the incident wave that is transmitted into the rest of the system during the transient period. Note that the transmitted voltage waves superimpose onto the initial voltage across the capacitive elements and have opposite polarity to the initial voltage. As the wave travels along the cable it is attenuated due to the resistance of the conductor and screen. This is not considered in the above analysis. In addition, in the above analysis, it is assumed that a fault is located at an infinitely long distance along a cable. However, in reality the reflected waves travel back to the fault location and return but with reversed polarity to the converter station. The process continues until the traveling waves die out because of attenuation. This creates oscillating voltage and current waves at the DC bus to which the faulted cable is connected. This is illustrated via simulation results in the next subsections.

Although much depends on the system under consideration, the magnitude of L_{DC} is determined based on the following parameters (in addition to the system voltage),

1. the achievable fault neutralization time considering the worst-case relay time of the protection system and the maximum internal commutation time of the HVDC CB.

2. the maximum current withstand capability of the system components considering the total current breaking time – the fault current must not exceed this value.

3. the maximum current interruption capability of the HVDC CB – the CB must be able to interrupt any short-circuit current resulting in the system. Thus, the maximum current interruption capability must be coordinated with the maximum allowable short-circuit current in the system with sufficient margin.

4. preferably the converters connected to the healthy lines should be able to continue their controlled operation. Thus, the DC voltage of the healthy part should remain within the acceptable limit for continued controlled operation.

4.3.2. MMC converter DC fault response

In this section, the events subsequent to the arrival of a traveling wave at a converter station are described. Referring to Figure 4.3a, three distinct temporal stages; namely, SM capacitor discharge stage (green dashed line), arm current decay stage (blue dashed line) and AC in-feed stage (red dashed line) can be identified during a fault condition.

4.3.2.1. Sub-module capacitor discharge stage ($t_1 < t < t_2$)

The part of the incident voltage wave that is transmitted to the MMC through L_{DC} propagates into the three phase-legs of the converter. Here, the wave partly discharges the SM capacitors, resulting in a rapidly increasing current. Figure 4.4 shows the converter output AC voltage (top graph) and converter transformer primary (grid) side AC voltage (bottom graph). Since the wave travels back and forth between the fault point and the MMC, it causes identical oscillations in each of the phase legs as shown in Figure 4.4a from t_1 onwards. Nevertheless, the line-to-line voltage remains unaffected since the oscillations in the phases cancel out. This can be seen from Figure 4.4b, which depicts the line-to-line voltages on the AC grid side of the converter transformer. The converter keeps control

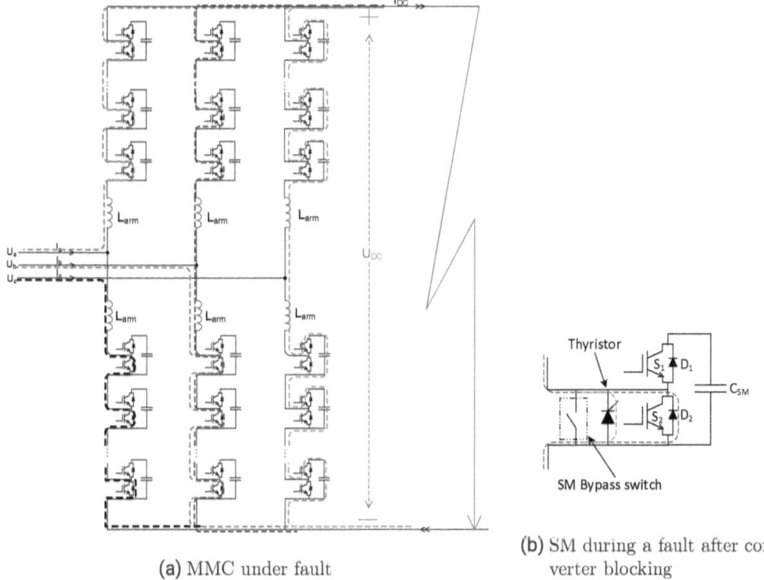

(a) MMC under fault

(b) SM during a fault after converter blocking

Figure 4.3.: Short-circuit current paths in an MMC at different stages after a fault occurrence

of its AC side parameters (AC currents and line-to-line AC voltages) until one of the criteria for converter blocking is reached – at t_2 in this case.

Therefore, a converter is in full control of its AC voltage until it blocks at t_2. The flattened peaks of the phase voltages in Figure 4.4a are due to third harmonic injection from the converter control. Normally, sufficient number of SMs per arm must be inserted to produce the AC peak voltage. However, by adding a third harmonic signal to the reference voltage generated by upper level control the converter arm no longer needs to produce peak AC voltage, rather a slightly bulged but reduced peak AC voltage [Cig14, SDM+14]. This increases the power rating of a converter without increasing the number of SMs – increasing the amplitude modulation ratio (index)[7] up to 1.155

---

[7]ratio of AC voltage peak value to DC pole voltage which is normally a value between zero and unity

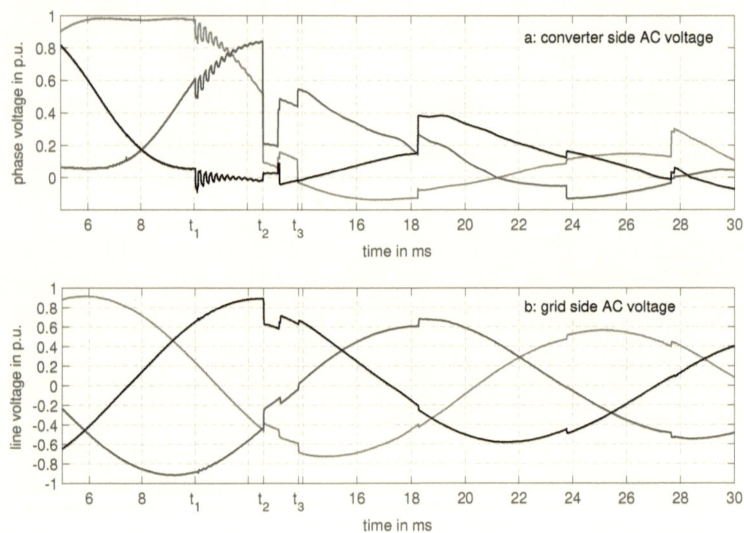

Figure 4.4.: a: Secondary (converter side) voltage. b: primary (grid side) voltage of a converter transformer. ©[2018]IEEE.

[SHN$^+$16]. Similar to the oscillating traveling waves, the line-to-line voltage remains unaffected since the artificially added third harmonics cancel each other due to transformer secondary side delta connection.

L_{DC} reduces the discharge of SM capacitors of the MMC by limiting the magnitude of the transient voltage wave transmitted into the converter. Therefore, depending on the inductance of this reactor it can significantly delay converter blocking. Figure 4.5 shows the current through the converter arms along with its AC output currents (per unitized with peak value of converter AC current). Although of very small amplitude, a similar oscillation as the converter AC voltage shown in Figure 4.4a can be observed. As for the AC voltage, it can be seen from the figure that the converter can keep control of its AC currents although the DC components of the arm currents are increasing. The increase

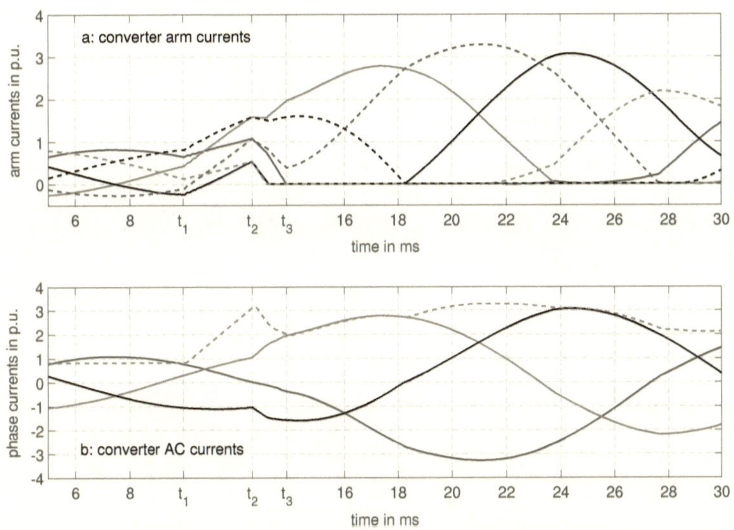

Figure 4.5.: a: arm currents (solid lines for upper arm and dashed lines for lower arm current). b: converter three phase AC currents (solid lines) and converter DC current (dashed line). ©[2018]IEEE.

in the DC components of the arm currents is due to the additional discharge of the SM capacitors caused by the fault. Theoretically, the discharge of the SM capacitors continues until the converter is no longer able to control its AC and/or DC outputs. However, in practice, the converter blocks before it gets to this situation. The converter blocks either to protect its circuitry against overcurrent or when the minimum threshold value of SM capacitor voltage[8] is reached. By blocking the converter, further discharge from the SM capacitors is prevented.

The DC current of a converter is shown by the dashed magenta trace in Figure 4.5b. The rate-of-rise of discharge current from the SM capacitors is limited not only by L_{DC}

---

[8]the value too low to supply control boards and gate drivers

but also by the arm reactors (L_{arm}) as well as by the inherent inductance of the line up to the fault location. During this period, each phase leg can be considered as an RLC circuit operating in parallel (resulting in the equivalent circuit shown in Figure 4.6). Furthermore, it is assumed that the capacitor (voltage) balancing algorithm (CBA) is in action. This means all the capacitors in one phase leg ($2 \times N_{\text{SM}}$, where N_{SM} is the number of SMs per phase arm) will be discharged although only N_{SM} per leg are inserted at a time. Therefore, it is reasonable to assume that two sets of capacitors exist as if connected in parallel per phase since the two sets discharge at different time instants considering the switching of the respective SMs. Therefore, in total $6 \times C_{\text{eq}}$, where $C_{\text{eq}} = C_{\text{SM}}/N_{\text{SM}}$, discharge during the capacitor discharge stage. Figure 4.6 shows a simplified diagram of a converter during the SM capacitor discharge period.

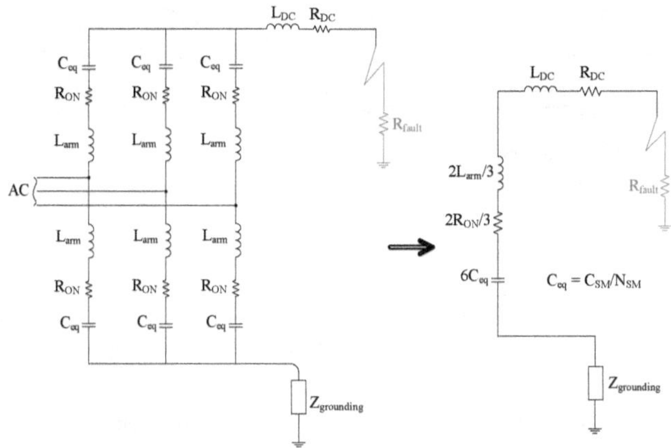

Figure 4.6.: Equivalent circuit of a converter during SM capacitors discharge period. ©[2018]IEEE.

Hence, assuming equal discharge of SM capacitors in the phase leg due to the CBA, the rate-of-rise of fault current from the converter can be obtained as (valid for a fault close to a converter terminal, until blocking),

$$\frac{di}{dt} = \left[B \left(\omega^2 + \alpha^2 \right) \sin\left(\omega t \right) - A \left(\omega^2 - \alpha^2 \right) \cos\left(\omega t \right) \right] \exp(-\alpha t) \tag{4.7}$$

where, $\alpha = \frac{R}{2L}$, $\omega = \sqrt{\frac{1}{LC} - \left(\frac{R}{2L}\right)^2}$, $A = CU_{\text{DC}}$, $B = \frac{I_0 + A\alpha}{\omega}$, $R = R_{\text{DC}} + R_{\text{fault}} + \frac{2}{3}R_{\text{ON}}$, $L = L_{\text{DC}} + L_{\text{line}} + \frac{2}{3}L_{\text{arm}}$, $C = 6C_{\text{eq}}$, I_0 is the initial steady state DC current. Neglecting the resistance in the circuit, 4.7 can be simplified further,

$$\frac{\mathrm{d}i}{\mathrm{d}t} = \frac{U_{\text{DC}}}{L} \cos\left(\omega t\right) \tag{4.8}$$

From Equation (4.7), it follows that the capacitance of the SM capacitor has an impact on the rate-of-rise of current from a converter. The capacitance of the SM capacitors are chosen such that the voltage ripple across these capacitors remains within tolerable limits during steady state operation, for instance, within a maximum of ± 10 % has been suggested as acceptable [JKA$^+$10]. For small SM capacitors, the under-voltage threshold for blocking is reached before the overcurrent threshold for the arm current is reached. The value of the SM capacitor is proportional to the converter rated power [Cig14].

Another important point that is often overlooked is that, during this stage, the MMC is normally exchanging power with the AC side. That means the SM capacitors are not only discharged into the DC fault but also recharged from the AC side. The best way to take this effect into account is to assume continuous I_{DC} in addition to the capacitor discharge obtained from the solution of Equation (4.7).

4.3.2.2. Arm current decay stage ($t_2 < t < t_3$)

This is an intermediate stage that results from the energy stored in the arm reactors during the capacitor discharge period. During a DC fault, when a converter has not yet blocked, the DC components of the arm currents increase without affecting the phase currents (in the interval $t_1 - t_2$ in Figure 4.5). When the converter blocks, the arm currents cannot be instantly switched off because of the arm reactors. In fact, by blocking the converter the SM capacitors are bypassed, and there is no further discharge from these capacitors – see the dashed blue line in Figure 4.3a. However, the arm currents continue to flow through the freewheeling diodes (from t_2-t_3 in Figure 4.5), and since the currents in all the arms have positive values, the freewheeling diodes in all the arms are conducting. Thus, for some time, the AC side is essentially decoupled from the DC side. Besides, the DC voltage of the converter drops significantly during this period since at that time, there is no inherent voltage source supplying the DC side. This is illustrated in Figure 4.9b in Section 4.3.3.

On the AC side, until one of the phase arms ceases to conduct, the fault appears to be a three phase fault with an AC impedance equal to,

$$Z_{total} = Z_{AC} + Z_{Tr} + \frac{1}{2}Z_{arm} \qquad (4.9)$$

Where, Z_{AC}, Z_{Tr} and Z_{arm} are, respectively, the equivalent impedance of the AC system at the point of common coupling (PCC), the converter transformer and converter arms.

However, due to resistance in the current path, the currents in three of the arms having the lowest values soon decay to zero, thus changing the converter into a diode rectifier mode of operation. This also changes the impedance seen by the AC side. Similarly, the DC current from a converter also decays with the arm currents until the AC in-feed starts as can be seen between t_2-t_3 in Figure 4.5b.

4.3.2.3. AC in-feed period ($t > t_3$)

After current through three of the arms decayed to zero (at t_3,) the DC current output from the converter is entirely comprised of AC in-feed through a three phase diode rectifier, see Figure 4.7. In an ideal diode rectifier only two arms, one from the top and the other from the bottom, are conducting. However, due to the inductance of the arm reactors and the transformer leakage reactance (including any AC side reactance) at least three of the phase arms are conducting at a time. In fact, later, if the fault is not cleared by HVDC CBs or if the AC side is not disconnected from the converter by AC CBs, there will be a commutation overlap between more than three arms as the AC in-feed current through each arm increases. Moreover, because of this commutation overlap, the DC voltage is not exactly a six-pulse rectified voltage. This is illustrated later in Figure 4.9b in Section 4.3.3.

It is important to note that the magnitude of the AC in-feed depends on the strength of the AC network at the PCC, the impedance of the converter transformer, arm reactors as well as the impedance on the DC side up to the fault location. The average DC current coming from the AC side can be described mathematically as follows [MUR02],

$$I_{DC} = \frac{\frac{3\sqrt{2}}{\pi}U_{AC}}{Z_{total}} \qquad (4.10)$$

where, U_{AC} is the AC line voltage on the secondary side of a converter transformer. The total impedance (Z_{total}) is given by,

$$Z_{total} = Z_{AC} + Z_{Tr} + \frac{1}{2}Z_{arm} + \frac{2}{3}\left(R_{DC} + R_{fault} + Z_{G}\right) \qquad (4.11)$$

Where, Z_{AC} is the short-circuit impedance of the AC system at the PCC transferred to the secondary side of the transformer, Z_{Tr} is the impedance of converter transformer, Z_{arm}

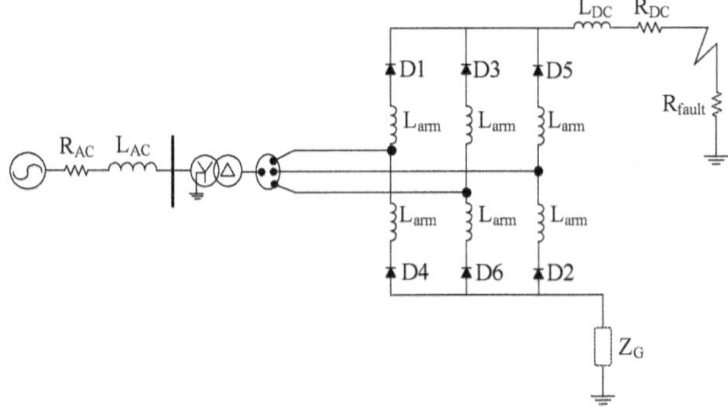

Figure 4.7.: Equivalent circuit diagram during AC in-feed stage of fault condition

is the arm impedance consisting of inductance and resistance of the arm reactor as well as total resistance of the N_{SM} series connected diodes, R_{DC} is the DC line resistance up to the fault location, R_{fault} is the fault resistance, and Z_G is the neutral point grounding impedance.

The above mathematical expression provides the AC current in-feed at steady state, and hence L_{DC} does not have an impact. Thus, the stronger the AC grid, the larger the AC in-feed current. Also note that the strength of the neighboring AC network, however, does not affect the fault current until a converter blocks.

4.3.3. Analysis of fault current contributions

If only one cable is connected to a DC bus, then the same current coming from the converter would flow through the faulted cable and the HVDC CB connected to it. In this situation the need for an HVDC CB is not mandatory as the fault can be cleared by an AC CB on the AC side of the converter. However, when there are multiple lines connected to the DC bus, as shown in Figure 4.8, the faulty line needs to be selectively isolated to reduce the impact on the healthy part of the DC grid; thus, HVDC CBs are required. Moreover, the presence of multiple connections at the DC bus has an impact on the fault current magnitude flowing through the HVDC CB of a faulted line.

To illustrate the impact of multiple incoming lines at the DC bus, simulations have been

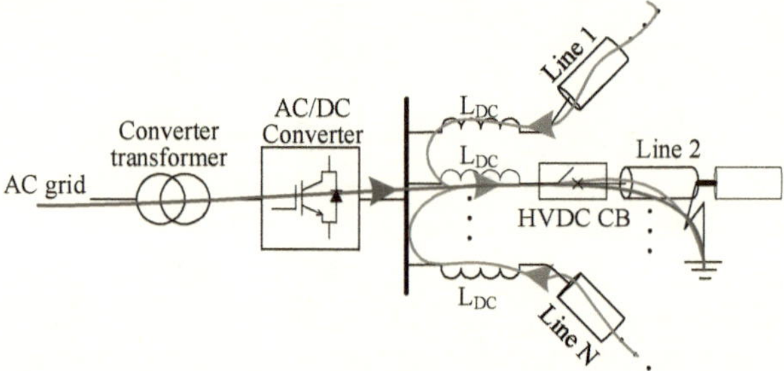

Figure 4.8.: DC fault close to a converter terminal with multiple connections. Fault current contributions from the AC side (blue) and DC side (red). ©[2018]IEEE.

performed. Figure 4.9 shows simulation results when a fault is applied close to a converter with three cables connected to its DC bus. Figure 4.9a depicts current oscillograms (per unitized with converter DC current); the current through the CB of the faulted cable (solid red curve), the current contributed by the converter side (dashed blue curve) and current from the discharge of two healthy cables (dotted black and dash-dotted magenta curves). Because of the additional discharge current from the adjacent cables, the rate-of-rise of current through a CB is higher than the rate-of-rise of current at the output of the converter. Nevertheless, the converter is the dominant contributor during the SM capacitor discharge stage (see between t_1-t_2 in Figure 4.9a).

Considering individual contributions, the discharge from the healthy cables is suppressed by two L_{DC}; one at the end of the cable itself and the other at end of the faulted cable – see Figure 4.8. This can be seen from Equation (4.6) where the transmitted voltage wave at the DC bus is further divided between the drop across L_{DC} and the part that propagates along the cable. Nevertheless, the current contributions of the healthy cables become significant after the converter blocks (from t_2 onward in Figure 4.9a). Especially, during the arm current decay period, the voltage at the DC bus drops significantly since there is no inherent voltage source as mentioned earlier. The collapse in the DC bus voltage during this period creates another transient resulting in further discharge from the adjacent cables and hence, the current through the HVDC CB keeps increasing although the current contributed by the converter is decreasing (see between t_2-t_3 in Figure 4.9a).

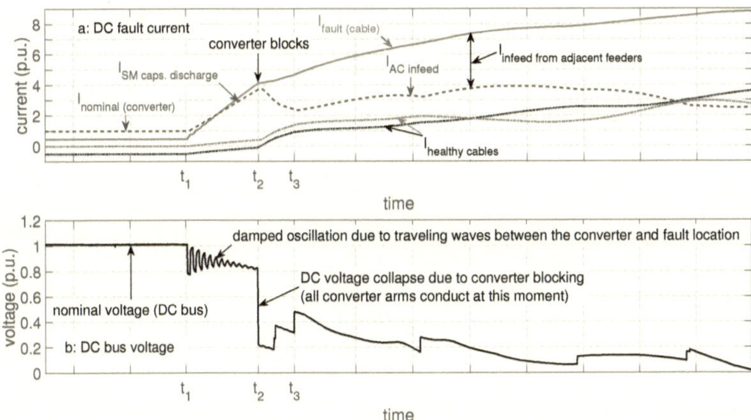

Figure 4.9.: a: Converter DC current (dashed blue) and current at CB location (solid red curve). b: Converter DC output voltage. ©[2018]IEEE.

4.3.3.1. Impact of fault location

One of the key factors affecting the rate-of-rise of fault current is the location of a fault within an MTDC grid. The location of the most severe fault condition is determined by several factors including the type of transmission line (an overhead line or a cable), the number of feeders at the bus to which a faulted link is connected, the distance of a fault from the DC bus, the power rating of a converter to which a faulted line is connected, the speed of the protection system including that of HVDC CB (fault neutralization time), etc.

Figure 4.10 shows simulation results of faults on a cable at different distances from a converter. For comparison purpose, the time delays of the traveling waves from the fault locations to the converter terminals are compensated in this figure adjusting the start of rise of the fault current to the same point in time in the graph. As the distance of a fault from a converter increases, the initial rate-of-rise of fault current increases. This can be seen from the curve of a fault at distance of 100 km in Figure 4.10a having highest initial rate-of-rise of current among the simulated cases. However, beyond a certain distance determined by the cable characteristics, the initial rate-of-rise of fault current starts to decrease due to attenuation of the traveling waves as they propagate through longer distances. This is the reason why the current for a fault at 100 km distance has higher initial rate-of-rise compared to the current for a fault at 240 km since the traveling waves are attenuated less in 100 km. Nevertheless, this is valid only until the reflected traveling waves propagate to the fault location and come back to the converter terminal. Hence, assuming a fault neutralization time of T_{FN}, the time from occurrence of a fault until a HVDC CB builds a TIV for current suppression, the highest theoretical rate-of-rise of fault current is when a fault occurs at a distance of at least S, where S is determined as in [SR16, CBS15],

$$S = \frac{vT_{FN}}{3} \tag{4.12}$$

Where v is the speed of the traveling wave through the cable. Note that T_{FN} is measured from the moment the fault occurs and not from the moment the current starts to rise at the location of the HVDC CB. Equation (4.12) is valid assuming zero attenuation of traveling waves in a cable. However, practically the resistance up to the fault location (including fault resistance) significantly attenuates the traveling waves and thus, reduces the rate-of-rise of current.

Furthermore, considering the AC in-feed period, the longer the distance of a fault to a converter, the higher the DC side impedance. Therefore, during the AC in-feed period, a fault at a converter terminal results in larger AC in-feed current. This can be observed from Figure 4.10a where a fault at converter terminal has the highest magnitude after t_2. Moreover, assuming the same L_{DC} is used, a fault at a converter terminal results in a large current for over-head transmission lines (OHLs) interconnected systems. This is because of the inductive nature of OHLs which limits the rate-of-rise of current. In general, considering achievable relay time, which is up to 3 ms, and the state-of-the-art HVDC CBs, which require operation time of at least 3 ms, a fault at a converter terminal for any case results in the highest average rate-of-rise of current.

4.3.3.2. Impact of system grounding

System grounding has an impact on the fault current magnitude and/or rate-of-rise especially during a pole-to-ground fault in, for example, a bipole system. Depending on the type, location and impedance, the grounding can limit the rate-of-rise of fault current in case of inductive grounding or limit the peak value of fault current in case of resistive grounding. Different types of grounding schemes can be used in HVDC systems, which have been discussed in [LTBH14]. In fact each type of grounding has an impact on the neutral bus voltage. Also, depending on the converter configuration, it can be

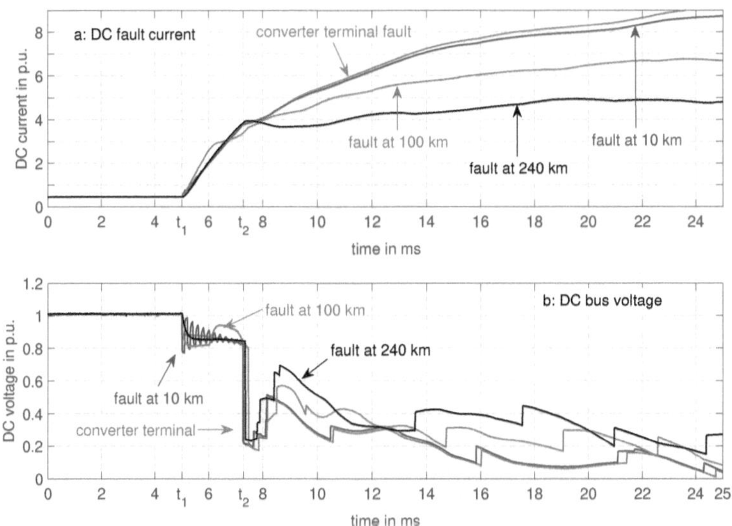

Figure 4.10.: Impact of distance of fault location from a converter. a: Fault current b: DC bus voltage. ©[2018]IEEE.

implemented either on the DC side or on the AC side [Cig17b].

Nevertheless, the impact of grounding is negligible during a pole-to-pole or pole-to-neutral fault.

4.3.4. Impact of converter blocking

In point-to-point VSC systems, a converter blocks (following a fault) to prevent damages to its components due to overcurrent caused by the excessive discharge of its SM capacitors. It might also block because it is unable to continue its controlled operation due to the converter voltage falling below a certain threshold[9].

---

[9]This is essential because the converter electronic circuits (such as the PE gate units) could be supplied from the SM capacitors [IEC]

In an MTDC grid where selective protection is realized using HVDC CBs, the need and the criteria for converter blocking are rather unknown territory and a closely guarded secret by the manufacturers. First of all, whether a converter has to block or not under such operational condition is not clear yet. Even if a converter blocks, what are the criteria for blocking and how is it coordinated with the operation of HVDC CBs? Also, what are the criteria for de-blocking[10] assuming a fault is cleared by HVDC CB? These criteria require extensive system studies, which are beyond the scope of this book.

Nevertheless, a converter blocking has a significant impact on a system as well as on an HVDC CB. The impact of a converter blocking during a fault situation on the stresses seen by an HVDC CB is discussed in Chapter 5. Regarding the impact on a system, simulation results show that converter blocking accelerates system DC voltage collapse. Figure 4.11 depicts converters' DC bus voltages under the two operation conditions. The dashed lines show converters' DC bus voltage profiles under the assumption that

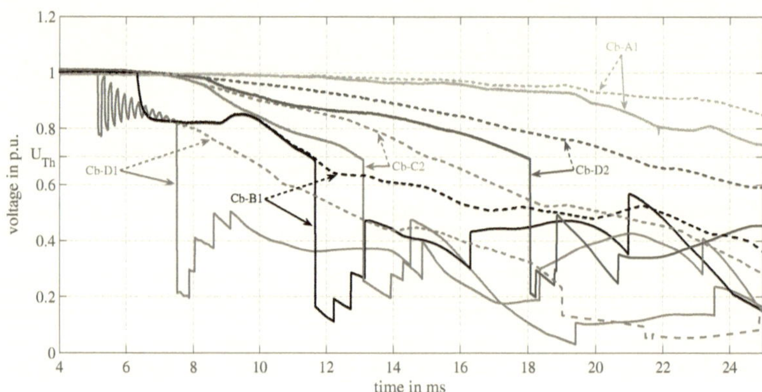

Figure 4.11.: Simulation result showing the impact of converter blocking on DC bus voltages

converters remain de-blocked[11] (even if a fault is not removed) whereas the solid lines show converters' DC bus voltages when converters block during a fault condition. In the

---

[10]for how long can a converter remain blocked before having to recharge SM capacitors?

[11]condition of the converter, in which turn-on and turn-off signals are applied repetitively to IGBTs of the converter [IEC]

latter case, a converter blocks when any of the arm currents exceed 3.5 kA and/or when the DC bus voltage falls below 0.7 p.u[12]. As indicated on the graph, the voltages of the same converter under the two operation conditions are plotted with the same color but different line styles.

Figure 4.11 shows that, as soon as a fault is applied, converter C-D1 blocks first due to its proximity to a fault location. This converter blocks because of overcurrent since the under-voltage threshold (U_{Th}, shown on the graph) is not reached. A clear observation from the figure is that the blocking of C-D1 initiates a more rapid drop in the bus voltage profiles of the other converters, most notably C-C2 and C-D2 which are located close to C-D1. This is because the voltage drop as a result of C-D1 blocking initiates a new transient condition in the system which enhances the voltage collapse of the other converters. Considering the points (on the graph) at which these converters block, it can be seen that all converters except C-D1 blocked because of the under-voltage threshold (U_{Th} in the figure). Since the converters C-D1 and C-B1 are situated on the either ends of a faulted cable, the impact of the former converter blocking cannot be seen on the latter. When a fault is cleared by the HVDC CB, the system restores before the threshold for blocking in all converters except C-D1 is reached.

4.4. Summary and Conclusion

In this chapter, a fault condition in a conceptual MTDC grid and the main contributors to the fault current as well as the critical factors determining the parameters of the fault current (rate-of-rise and magnitude) are investigated. The major events during fault current build-up are discussed in three distinct stages. During the SM capacitor discharge stage, the fault current is mainly supplied by the converter directly connected to the DC bus at the end of the faulted line. During the arm current decay period, the discharge from other feeders connected to the DC bus becomes prominent. The strength of the AC grid and the impedance of the converter transformer become critical during the AC in-feed stage. The overall current through a CB on a faulted cable increases with the number of cables connected to the DC bus. It also concluded that considering the breaker operation time of the recently developed HVDC CBs, a fault close to (at) a converter terminal with multiple connections results in the worst-case fault current. Proper size of L_{DC} can reduce the fault current to remain within interruption capability of the CB.

---

[12]both thresholds chosen in this book

5. Fault Current Interruption in MTDC Grids and Stresses on HVDC CB

Considering the stringent requirements of the VSC HVDC transmission schemes, several manufacturers have proposed and developed prototypes of HVDC CBs as discussed in Chapter 3. The functionality and performance of these prototypes have been studied and demonstrated through a range of development tests in the manufacturers' own labs. However, meaningful demonstration is achieved only when tests accurately reflect realistic fault conditions. For this purpose, this chapter focuses on simulation study of fault current interruption process in a MTDC grid, and analysis of the resulting electrical stresses on the HVDC CBs. In Section 5.1 fault current interruption process by active current injection HVDC CB and the resulting stresses are discussed. The impact of converter blocking on the stresses of an HVDC CB is presented in Section 5.2. The important electrical stresses on HVDC CB and the critical stages of fault current interruption are summarized in Section 5.3. Finally, the summary and conclusion based on the results of the study in the chapter are provided in Section 5.5.

5.1. HVDC Fault Current Interruption – using HVDC CB

Nowadays several different concepts of HVDC CB are developed for application in MTDC grids. The two leading technologies are hybrid- and active current injection HVDC CBs, several realizations of which are discussed in Chapter 3. In all cases the fault current breaking strategy involves local current interruption in the CCB followed by internal current commutation to the branch(es), which generate the TIV and ultimately current commutation into the EAB.

Simplified electrical models of different HVDC CBs have been developed considering realistic parameters as recently demonstrated. The models are embedded into the benchmark MTDC grid to realize a fully selective fault clearing strategy i.e., at the ends of each cable as shown in Figure 4.1, to investigate interactions of these CBs with the system.

Figure 5.1 shows the current interruption process by a direct capacitor discharge based active current injection HVDC CB (described in subsection 3.3.2) along with the associated stresses during the current interruption. Under normal operation, the capacitor of the CB is pre-charged to the system voltage (320 kV). It is assumed that local current interruption in the CCB is achieved upon the first current zero crossing although it could happen at later current zeros. The impact of the latter is discussed in Chapter 9. A fault neutralization time of 9 ms, of which 2 ms is relay time, is assumed in the simulation. In order to limit the fault current to less than 16 kA over this period a 150 mH L_{DC} is placed in series with the CBs.

Considering Figure 5.1a, a pole-to-ground fault is applied at t_0, which is instantly followed by a steep rise of the current. The dashed black and red traces in the figure, respectively, show the resulting current and voltage when no action is taken by the HVDC CB and/or the protection system. The traces in solid lines show simulation results when the HVDC CB successfully clears the fault current. It is assumed that a protection system detects and locates the fault between t_0 and t_1. Hence, at t_1 a trip command is sent to the HVDC CBs on the faulted cable. This marks the beginning of the breaker operation time. At t_2 the converter blocks and at t_3 the AC in-feed takes over. At t_4 the breaker is ready to withstand the TIV and thus, a counter current is injected from the pre-charged capacitor to a create local current zero in the CCB. Then current commutates into the commutation branch charging the capacitor until the protection level of the EAB is reached. The latter leads to the fault current commutation to the EAB, which finally suppresses the fault current until it reaches a leakage level at t_6.

For any kind of HVDC CB, the EAB consists of several parallel stacks of metal oxide surge arresters (MOSA) designed to clamp the peak TIV as well as to absorb the magnetic energy stored in the system inductance. Figure 5.1b shows the total energy absorbed by the surge arrester of the active current injection CB. Depending on the size of L_{DC}, the magnitude of the interrupted fault current and the system voltage during interruption, up to several tens of MJ of energy needs to be absorbed. For the simulated case, energy of about 18.5 MJ is absorbed by the CB while interrupting 11 kA.

The total energy absorbed is further decomposed into two sources; namely, the magnetic energy initially stored in the series L_{DC} at the beginning of fault current suppression and the electrical energy coming from the remaining part of the system during the fault current suppression. Note that the energy stored in L_{DC} at the beginning of the energy dissipation phase is $\frac{1}{2}LI_p^2$, and in the Figure 5.1b, the curve in black depicts the dissipation of this energy during the fault current suppression period. The total energy that a CB must absorb is given as,

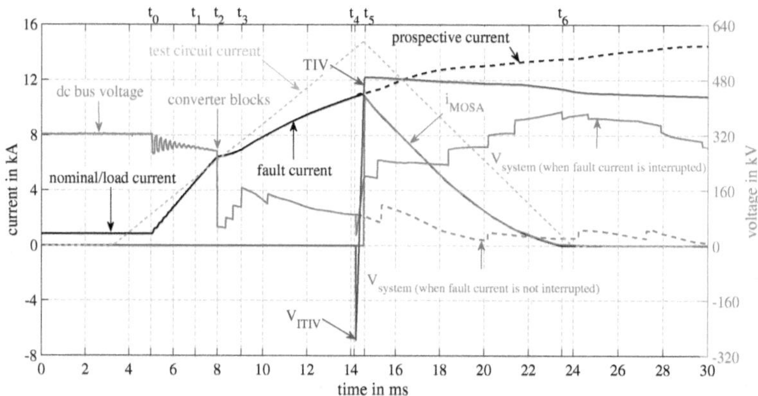

(a) Prospective fault current (dashed black) and the corresponding system voltage (dashed red) when breaker takes no action; current through (solid black), voltage across the CB (solid blue), system voltage (solid red), MOSA current (solid pink) during current interruption. The dashed green trace shows an envelope current for testing the HVDC CB. At t_0 fault current start to rise, at t_1 the breaker receives trip command, at t_2 converter blocks, at t_3 AC in-feed starts, at t_4 counter current injection, at t_4–t_5 TIV generation, at t_6 fault current suppressed

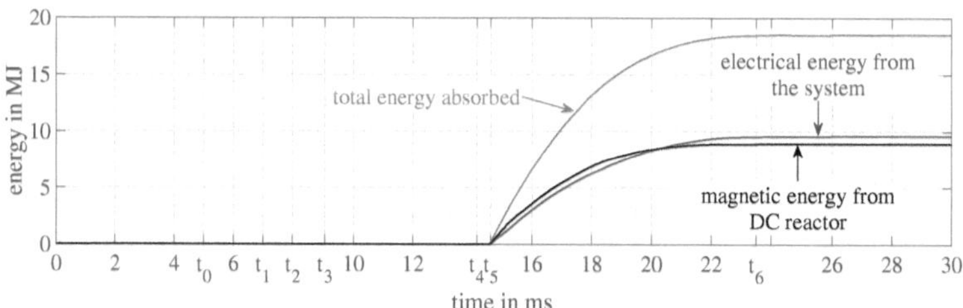

(b) energy absorbed by CB – decomposed into magnetic energy and electrical energy

Figure 5.1.: Stresses on active current injection HVDC CB. ©[2018]IEEE.

$$E_{total} = \frac{1}{2}L_{\mathrm{DC}}I_{\mathrm{p}}^2 + \int_{t_5}^{t_6} U_{\mathrm{DC}}i_{\mathrm{MOSA}}\mathrm{d}t \qquad (5.1)$$

Where I_{p} is the peak value of the interrupted current and, i_{MOSA} and U_{DC} are current through the CB and the recovering DC bus voltage, respectively, during energy absorption period ($t_5 - t_6$ in Figure 5.1a).

A noteworthy difference between AC and DC current interruption is that in the latter case the system voltage starts to recover not at the end of the current interruption process but rather from the moment the CB generates the counter voltage (from t_5 in Figure 5.1a). That is while the fault current is still at its peak. As a result of this, there is significant electrical energy coming from the system during the current suppression period.

Thus, it is not only the system magnetic energy that the CB must absorb but also the electrical energy supplied by the system during the fault current suppression period. The latter comes inevitably since the breaker requires some time to dissipate the magnetic energy in the reactor, with the duration depending on the value of L_{DC} as well as the difference between the TIV and the DC bus voltage [BSTI16]. Due to the system voltage recovery during the energy dissipation phase, the energy coming from the grid in this case is even larger than the energy stored in L_{DC} (see Figure 5.1b).

5.2. Impact of Converter Blocking on the Stresses of HVDC CB

The other impact of a converter blocking is on the stresses seen by an HVDC CB in terms of maximum breaking current, TIV duration as well as energy absorption. Because the SM capacitors continue to discharge when a converter remains de-blocked, the fault current rises to a higher peak than when a converter blocks. This can be seen from simulation results shown in Figure 5.2a where the prospective fault current under the two converter operation conditions is depicted. Hence, when an HVDC CB is employed to clear a fault, it interrupts different current levels under the two operation conditions.

A model of a hybrid HVDC CB, implemented in PSCAD/EMTDC, is embedded into the benchmark grid to clear a fault assuming a relay time of 3 ms and a breaker operation time of 3 ms – a fault neutralization time of 6 ms. It can be seen from the Figure 5.2a that the breaker interrupts a lower current peak (12 kA) under blocked converter operation than when a converter continues de-blocked operation (15.5 kA). A noteworthy point is the duration of the fault current suppression under the two operation conditions. Although lower current is interrupted under blocked converter operation, the duration of the fault current suppression, and consequently the duration of the TIV stress, in this case is slightly longer than when a converter remains de-blocked. This can be seen from

Figure 5.2b. Nevertheless, the rate-of-rise and the peak of the TIV are more or less similar for the two situations.

As mentioned in the preceding subsection the system voltage starts to recover from the moment the HVDC CB generates the TIV. However, the voltage recovery for the two converter operation conditions differs considerably as can be seen from Figure 5.2b. When a converter blocks, the DC bus voltage becomes the output voltage of an uncontrolled rectifier. The DC voltage of an uncontrolled rectifier depends on the AC voltage, AC impedance, arm reactors as well as on the DC current. Although the relationship is much more complex (due to multiple commutation overlaps), the DC output voltage of a rectifier during the current suppression period can be approximated as follows [MUR02],

$$U_{\text{DC,rectifier}} = \frac{3\sqrt{(2)}U_{\text{AC,l-l}}}{\pi} - \frac{3}{\pi}X_{\text{total}}I_{\text{DC}} \tag{5.2}$$

where, $U_{\text{AC,l-l}}$ is the RMS value of the line-to-line AC voltage, X_{total} is the total AC side equivalent reactance including the leakage reactance of a converter transformer and arm reactance.

Equation (5.2) shows that as the fault current is suppressed, the output voltage of the rectifier increases. It is the uncontrolled recovery of the rectifier voltage which is the main cause for the increased duration of the TIV even if a lower current is interrupted. For the de-blocked converter operation the DC voltage is maintained by the insertion of SM capacitors. During the current suppression period, a converter attempts to regain its nominal DC voltage. For this, the SM capacitors need to be recharged.

The energy absorbed by the HVDC CB is higher when the converter remains de-blocked – due to the higher current interruption. However, the recovering system voltage also affects the amount of electrical energy contributed as per Equation (2.11) and (5.1). Figure 5.3 shows the energy absorbed by the HVDC CB when interrupting fault currents as shown in the Figure 5.2a. Under blocked converter operation, the CB absorbs about 20 MJ energy of which only 35% comes from magnetic energy stored in the series reactor. The remaining 65% is contributed by the recovering system voltage. However, in the case of de-blocked converter, the HVDC CB absorbs 25 MJ energy. In this case L_{DC} contributes about 61% of the total energy. An important conclusion is that it is not only the peak value of the current that determines the energy absorbed by the HVDC CB but also the trajectory of the recovering system voltage during current suppression. This must be carefully considered when defining the requirements of the HVDC CBs.

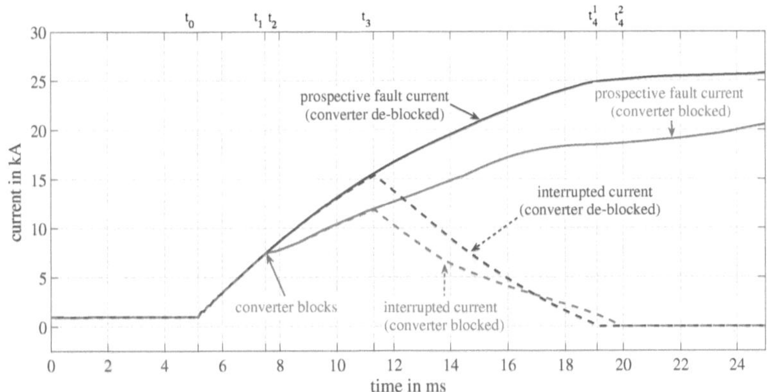

(a) Prospective and interrupted fault currents when a converter blocks and remains de-blocked during a fault

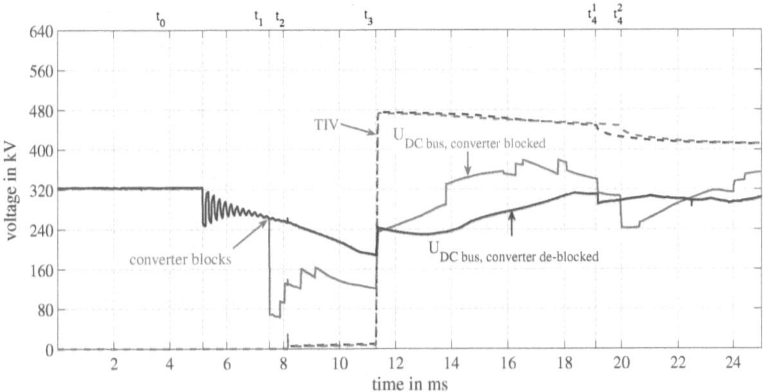

(b) Converter DC bus voltage during fault current interruption and TIV across CB

Figure 5.2.: Simulation result showing the impact of converter blocking on fault current and DC bus voltage

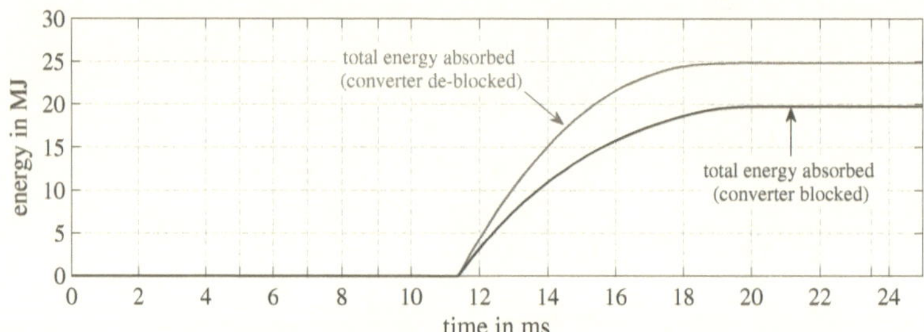

Figure 5.3.: Simulation result showing the impact of converter blocking on energy absorbed by HVDC CB

5.3. Stresses and Critical Stages of DC Interruption

Summarizing the discussion in the previous section, the basic electrical stresses on an HVDC CB that need to be reproduced during a short-circuit current interruption test are the following:

1. **Current stresses** – the maximum breaking value with sufficient rate-of-rise. Moreover, the trajectory of current is essential, for example, for the thermal stresses of arcing mechanical gaps or on the PE switches. The rate-of-rise shall be determined based on the worst-case relay time and internal current commutation time so that the test current envelopes all the possible short-circuit currents in the system.

2. **Voltage stresses** – this includes the TIV (also ITIV in some cases) during the dynamic/opening process and post current suppression DC recovery voltage that appears after the breaker is fully open.

3. **Energy stresses** – this is crucial stress which leads to thermal stress on the EAB of the HVDC CB while at the same time the mechanical and/or the PE components in the other branches are stressed by the sustained TIV during the current suppression period as mentioned in 2.

The key critical stages and the essential performance parameters at each stage that need to be demonstrated by an HVDC CB during a current interruption test are:

1. **Internal current commutation** – capability to create a local current zero without restrike/breakdown of mechanical switches/interrupters or thermal overload of PE components at the rated DC fault current interruption capability. The key design parameters (performance indicators from testing perspective) at this stage are:

   a) **Internal current commutation time** – the time from trip command until the peak value of interruption current is reached.

   b) **Maximum breaking current** – the maximum short-circuit current that the breaker can interrupt within the internal current commutation time. Compared to the maximum fault current of a system that can be seen at the end of fault current neutralization period, the maximum breaking current capability with sufficient margin is necessary. Moreover, the test current trajectory must encompass the system fault current excursions to ensure sufficient thermal stresses to the internal (sub-)components.

2. **Generation and maintenance of TIV** – sufficient rate-of-rise and magnitude to initiate fault current suppression. Although determined by the CB itself, the rate-of-rise, the peak and the duration of the TIV are crucial parameters during a test.

3. **Energy absorption** – capability of energy absorption components to withstand thermal as well as dielectric voltage stress during fault current suppression. Depending on the rated sequence, this capability must be demonstrated several times within a defined sequence.

4. **DC recovery voltage withstand** – after current suppression the breaker must withstand the rated maximum continuous DC voltage for a certain duration – for example, 300 ms.

The number of current breaking operations that the CB can perform before thermal degradation/damage occurs to its MOSA as well as the interruption intervals need to be defined, e.g. like auto re-closure in AC CBs. In addition, test duties covering the worst-case stresses expected in service must be defined. This is discussed in Chapter 8.

Moreover, the HVDC CBs are typically realized by connecting several modules in series to achieve high-voltage rating. Interruption tests can be performed on a single module or multiple modules. When performing tests on a single module, equivalent stresses that would appear across a single module, which forms part of full-pole HVDC CB, must be considered. In some cases, for example for HVDC CBs having PE switches in the CCB, pre-conditioning[1] to mimic worst case normal service conditions might be necessary.

---

[1]especially any thermal effects due to sustained load current flow prior to current breaking process

5.4. Requirements of Test Circuits

In reference to the stresses and the critical current interruption stages discussed in Section 5.3, various test circuits are evaluated in [BSTI16, BS17]. It is shown that besides producing sufficient short-circuit current, a test circuit having adequate voltage supply during energy absorption phase of the current interruption process is essential. The reason for having adequate voltage supply during this period is to mimic the electrical energy coming from the system, due to the recovery of a system voltage, during current suppression. Moreover, when the faulty line is isolated by CBs, the remaining part of the system continues to operate normally. Hence, after current suppression, the HVDC CB is subjected to DC recovery voltage stress from the system.

In general, a test circuit must be able to provide the most onerous stresses expected in a service condition. The most onerous conditions for HVDC CBs are expressed in terms of current, voltage and energy. In order to adequately supply these stresses, a test circuit needs to fulfill the following conditions:

1. **Produce sufficient short-circuit current that has adequate rate-of-rise** – within the fault neutralization time (considering the worst-case relay time and breaker operation time), current must rise to at least the peak current breaking capability of an HVDC CB. A test current must encompass the possible excursions of current during any fault in the system so as to consider the thermal effect on the internal components. Because the performance of some technologies of HVDC CBs can depend on the magnitude of the interrupted current, a test circuit must be able to produce a range of currents up to the peak current breaking capability.

2. **Supply sufficient energy as in service** – a test circuit needs to provide not only the energy equivalent to the magnetic energy stored in a system inductance during a fault but also the energy supplied by the (recovering) system during current suppression. Moreover, the energy must be dissipated in a time interval corresponding to a service conditions. That is, the rate of suppression of current and its time function must correspond to realistic conditions – ensuring sufficiently long duration of the self-imposed TIV stress on the breaker.

3. **Apply DC recovery voltage** – after successful current breaking by a CB, a test must ensure DC voltage stress equivalent to the maximum continuous rated voltage of a system for sufficiently long duration.

In order to fulfill the above requirements a test circuit needs to maintain its supply voltage not only up to the local current interruption in the CCB of the HVDC CB, but also during the entire energy dissipation phase of the current breaking process [BS17].

5.5. Summary and Conclusion

The electrical stresses on HVDC CBs: current, voltage and energy, at different stages of DC fault current interruption process, are identified based on simulation studies. Four critical stages: internal current commutation, generation and maintenance of TIV, energy absorption and DC recovery voltage withstand are identified along with essential CB parameters crucial from a system perspective.

During the current breaking process, the HVDC CB is stressed by the electrical energy supplied by the system in addition to the magnetic energy stored in L_{DC} during a fault. System simulation studies show that the electrical energy contributed by the system exceeds the magnetic energy stored in L_{DC}. Therefore, a test needs to produce sufficient short-circuit current with adequate rate-of-rise; that, within the worst-case fault neutralization, rises at least to the peak interruption capability of a CB. The maximum breaking current of the HVDC CB must envelope any possible fault current occurring in a system with sufficient margin. It is also necessary to supply sufficient energy stress as in service. For this a test needs to provide the energy equivalent to the energy stored in L_{DC} as well as the energy supplied by the rest of the system. Finally, after successful current interruption by a CB, a test must ensure sufficient DC voltage stress equivalent to the rated maximum continuous system voltage.

6. Review of Test Methods and Test Circuits for HVDC CBs

This chapter focuses on the test methods and circuits used for testing HVDC CBs in the literature. Most of its contents have been published in [BS17, BSTI16]. The chapter is organized as follows. Section 6.1 provides brief background of the subject while Section 6.2 briefly reviews some of the test methods used over the course of HVDC CB development. In Section 6.3 the actual test circuits used for performance demonstration of HVDC CB are discussed in detail. The advantages and the main limitations of the various candidate test circuits are described. Finally, the conclusions drawn based on the discussion of the chapter are presented in Section 6.4.

6.1. Introduction

Over the years the performance requirements of the HVDC CBs have been changing with the evolution of HVDC systems. For example, because of the current control capability of LCC converters, short-circuit currents in a system built from such converters are not much in excess of the normal load current [Puc68].

Nowadays the preferred converter technology for building MTDC grids is the MMC VSC. As discussed in Chapter 4, unlike the LCC systems, the half-bridge SM based MMC VSC systems do not have fault current control capability and as a result a significantly large current flow, with rapid rate-of-rise, during a fault. Thus, HVDC CBs combining very fast operation with large current clearing capability are required in such systems in order to avoid excessively large short-circuit currents. The most recently developed HVDC CBs, which can fulfill such requirements, are discussed in Chapter 3. On the other hand, the test circuits needed for demonstrating the performance of the latest developments of HVDC CBs must provide test currents equivalent to the fault currents appearing in MTDC systems. Hence, the requirements of test circuits for HVDC CBs have also been changing. As a result, the DC fault current breaking tests of HVDC CBs have become challenging because of the requirement of supplying a range of sufficiently high rate-of-rise of test current and subsequently a large energy stress [CIG78, BS17, BPS18b].

So far the challenge of testing the HVDC CBs has been twofold. Firstly, at the moment of writing this book, no international standards specifying test requirements exist. Secondly, no test circuit capable of supplying adequate and complete stresses to the HVDC CBs is used at this stage of development. In view of this, a CIGRE joint working group, JWG B4/A3.80 [CIGa], has been established to provide realistic guidelines regarding technical requirements, stresses and testing methods of HVDC CBs.

Strictly speaking, the testing of HVDC CBs requires a high-power DC source – supplying high-current and high-voltage simultaneously – which is not available at any test laboratory worldwide. It was clear from the early days of HVDC CB development that it is unlikely that such an expensive and elaborate test circuit will be set up for solely testing HVDC CB prototypes even if technically it makes sense [CIG78]. Thus, alternative test circuits providing equivalent stresses must be sought. With this regard, different kinds of test circuits have been used at different stages of HVDC CB development [CIG78, BS17]. Various approaches (test methods) have been considered to address some of the limitations of the test circuits or to meet some of the essential stresses that need to be replicated in a test. Nevertheless, for the above and many other reasons, the testing of HVDC CBs has been limited to, mainly, proof of concept of industrial prototypes.

6.2. Review of Test Methods

Depending on the purpose of a test and also on the available facility for testing, there are many different approaches to verify the functionality of HVDC CBs. The purposes of testing could be functional/operational testing of a component or current branch(es) or a complete module or full-scale HVDC CB. It could also be aiming at verifying parts of operation or sequence of operations or of a complete operation. It could be part of a development test, routine test or type test in which the rated performance of the HVDC CB is verified.

6.2.1. Component/Branch testing

Most of HVDC CB components are standard components used in non-standard applications and hence, face non-standard stresses. Before testing the performance of a complete HVDC CB unit, several step-by-step functionality and/or performance tests are conducted on various (sub-)components or (part of-) current branch(es). The mechanical switching gaps such as a UFD and vacuum interrupter(s) as well as PE switches such as IGBTs/BIGTs/IEGTs together with snubber circuits are tested separately before putting them together to form a complete HVDC CB [HHJ15, WHAD+15, WHC+16]. For example, IGBTs/BIGTs in the CCB are tested for current and voltage sharing, maximum current interruption, failure mode, etc. whereas the PEs switches in the commutation branch are tested for maximum current interruption and TIV withstand capability in [HJ11, CLBR12, SRH$^+$15].

The test circuits used for component testing are rather simple (mostly charged capacitors) and are designed to verify electrical and/or mechanical functionality/performance of the components in the different current branches [WLL$^+$19]. For example, a full-bridge SM[1] is tested for current interruption capability of up to 28 kA including verification of its control units. The objective is to subject a SM to the stresses it would see as part of complete HVDC CB. In this example case a SM is made to conduct current for about 3 ms before interruption after which a TIV of $\leq$ 4 kV is applied per IGBT [TZH$^+$].

While the functionality verifications of individual components are necessary steps in the product development, they cannot replace the overall performance test of an HVDC CB. It was clear from the early stages of HVDC CB developments that testing of components cannot be sufficient to prove the effectiveness of the HVDC CB [BHK74]. Tests must be performed in a circuit emulating the actual HVDC network where a complete HVDC CB is subjected to realistic stresses.

6.2.2. Modular testing – unit and multi-unit testing

Like HVDC converter valves, HVDC CBs are also designed in a modular approach to achieve a high-voltage rating [CIG78]. A module (breaker unit) of an HVDC CB is the smallest part that contains all the required functionality of the full-scale HVDC CB. In this regard, a step-by-step functionality test has been the most common approach employed before testing a full-scale HVDC CB [HJ11, TWZ$^+$, TZH$^+$, GDT$^+$18].

For HVDC converter valves, identical units (e.g., SMs) are placed in series in large numbers. In this case it is sufficient to test the functional performance of a section consisting some minimum number of SMs connected in series in order to validate the entire converter [IEC]. However, care must be taken when testing modular equipment like HVDC CBs comprising a number of different components and branches, which must operate in a prescribed sequence. Modular design of HVDC CBs is not necessarily a series connection of independent modules. As discussed in detail in Chapter 3, nowadays, different modular design approaches are considered for extra high-voltage ratings. The most common approaches are depicted in Figure 6.1.

Each of these approaches have been considered in the recently implemented HVDC CBs where each approach has its own technical pros and cons. For the modular designs depicted in Figure 6.1a and 6.1b, testing of a downscaled module cannot ensure rated

---

[1]which is a building block of some hybrid HVDC CBs, see Section 3.4.3

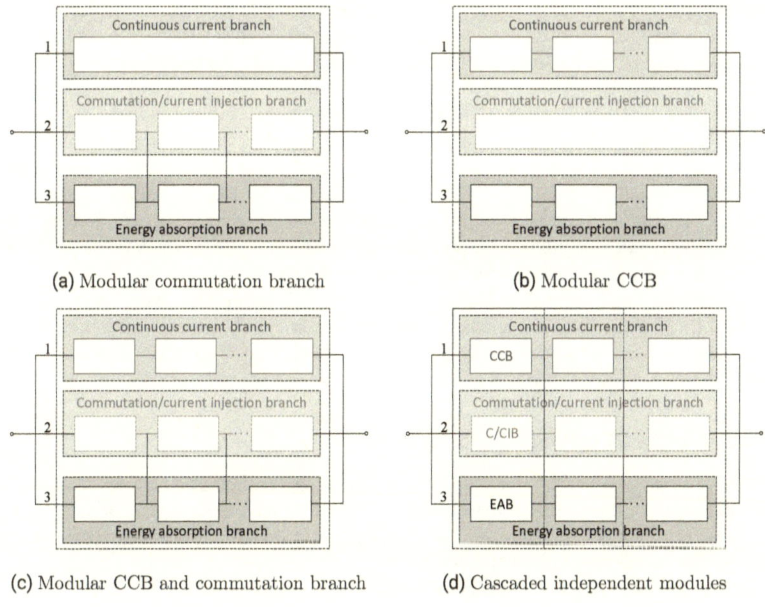

(a) Modular commutation branch   (b) Modular CCB

(c) Modular CCB and commutation branch   (d) Cascaded independent modules

Figure 6.1.: Different approaches of modularization of HVDC CB

stresses to every component since some current branches are naturally rated for full-pole voltage. Downscaled testing of such designs can only be used as a proof of concept [KHK⁺19]. The approach shown in Figure 6.1c has a modular design in each of its current branches. However, the modules are not necessarily independent in that the stresses seen by a module in one current branch do not necessarily correspond to the stresses appearing in a module of another current branch. Therefore, due attention must be given to the stresses applied during a test. A truly modular design, in which independent HVDC CB units are connected in cascade, is shown in Figure 6.1d. In this case a test of a single module can ensure application of the necessary stresses to each component of the different current branches. However, in such a design it is essential to ensure that the stresses seen by one module are exactly the same as the stresses seen by the other independent modules. In actual design, there might be some redundancy included to consider a failure of one module [LPM+20]. Therefore, it is important to extract the worst-case stresses seen by a module before testing. It might also be necessary to include some minimum number of modules in order to verify stress distribution. For example, before testing a full-pole 200 kV HVDC CB, 50 kV and 100 kV modules are tested in [TWZ+].

In any of the modular approaches shown in Figure 6.1, the components in a given current branch are subjected to the same current stress because of the cascade connection. What is different in each of the above design approaches are the voltage and energy stresses. The latter must be carefully downscaled considering the specific design. If an HVDC CB is built in a modular approach as shown in Figure 6.1d, the necessary stresses seen by a module can be extracted as shown in Figure 6.2.

Nevertheless, with careful stress extraction, modular testing remains an essential part of HVDC CB performance demonstration since no single equipment exists that fulfills extra-high voltage rating and/or no test facility exists that is capable of delivering the complete rated stresses to HVDC CBs above 400 kV system voltage [BPS19].

6.2.3. Multi-part testing

For large capacity AC CBs, tests are performed using synthetic test circuits because of the limitation of the short-circuit capacity of a test facility. A similar approach cannot be considered for HVDC CBs due to the fact that it is not only voltage and current equivalency that is essential for HVDC CB testing, but also energy equivalency is necessary [YTI+82]. At the time when energy of several mega-joules could not be supplied at any test facility, multi-part testing was suggested: one test of the CB (DC current interruption) and a separate test of the energy absorption devices. In some cases, separate tests have been carried out [EHL+76].

6.2.4. Field tests – staged short-circuit

The lack of sufficient energy at test facilities has led to field testing of HVDC CBs [HLR+76, MSLH76, VCP+85]. In the early days of HVDC system development, this was the only way to subject the HVDC CB not only to current, voltage and energy stresses but also to the actual electromagnetic interference (EMI) conditions as well as to the impact of traveling waves. Especially, the EMI generated in a converter station can be very high. Thus, exposing the HVDC CB controls to the actual operation can reveal the vulnerability of different designs [HLR+76, CIG78]. In addition, field tests accurately present the actual current and voltage excursions of the system during fault current interruptions, which is otherwise difficult to reproduce in a test circuit represented by

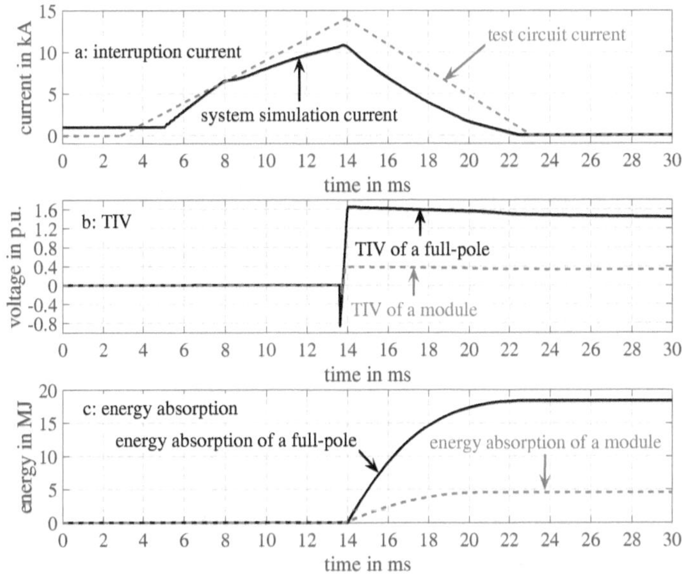

Figure 6.2.: Parameter (current, voltage and energy) downscaling for modular testing. Solid lines show parameters from system simulation considering a 320 kV CB. Dashed red lines indicate downscaled test parameters for one module rated for 80 kV.

lumped elements.

Although the field environment and operational stresses are desirable, execution of a test in such a setting is practically challenging. First of all such tests lack the flexibility of a test laboratory in that the test parameters are strictly limited to the system supporting the test. Also, to conduct such a test necessary arrangements must be made in advance so that minimum disturbance to the interconnected AC systems is insured. Needless to mention, the available time for testing is also limited. Coordination of operation of converters at different locations made this approach of testing simply inconvenient [HLR$^+$76].

Recently on-site tests have been performed in the MTDC pilot projects in China [LCQ18, Zha]. These tests are intended not only to verify the integrity of the HVDC CBs but also the coordination of system control and protection in the event of fault.

In fact field tests can be assisted with detailed and intensive computer simulations to identify the worst fault condition for the HVDC CB. Particularly, a great deal of system simulation is necessary to determine the energy absorption of HVDC CB and system overvoltage. In a simulation study a fault can be applied without HVDC CBs acting to check the prospective fault current and to investigate the impact of converter control on the fault current. An adequate model of a system that takes the dynamic response of the system control to the interaction between the system and HVDC CB into account is essential to rely on simulation results.

Nevertheless, in addition to qualifying the ratings and functional performance of HVDC CB, adequate laboratory testing can provide much needed information on the performance and withstand capability of HVDC CB [HLR$^+$76]. In a similar way as for AC CB, the use of artificial lines has been suggested to take the effect of traveling waves into account during DC CB testing [BHK74, CIG78].

6.3. Review of Test Circuits

Four types of test circuits supplied by different sources: namely, high-power rectifier, high-voltage charged capacitor bank, high-current charged reactor and variable frequency AC short-circuit generator are discussed in this section.

6.3.1. High-power rectifier circuit

The optimal test circuit is indeed a full-scale converter connected to short-circuit generators having sufficient power – providing full DC power and at the same time capable of withstanding the peak TIV. This incurs significant investment to build one at a test facility. Nevertheless, considering the system studies in Chapter 4 the role of the converter control during a short-circuit is rather limited. During a fault a converter is either blocked to protect its components or remains de-blocked relying on the protection system. These situations can be covered by a simple high-power Graetz rectifier bridge in combination with high-voltage capacitors, see Figure 6.3, without a need for full-scale converter along with its sophisticated control.

Assuming sufficient available short-circuit power for testing and proper circuit components, a test circuit based on a high-power rectifier circuit is ideally suited to replicate

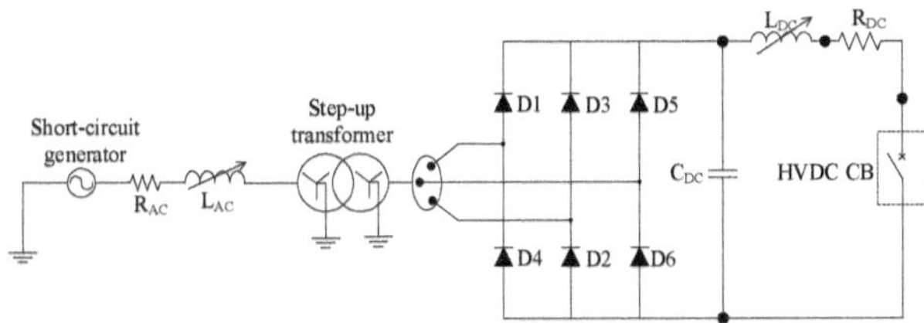

Figure 6.3.: Test circuit based on high-power rectifier supplied by short-circuit generators

the short-circuit current from a converter. A high-voltage capacitor (C_{DC}) is needed to maintain the DC output voltage and supply additional short-circuit current. The larger this capacitor, the better the test circuit with respect to supplying stresses to an HVDC CB. Additionally, an L_{DC} is placed in series with the HVDC CB to limit the rate-of-rise of current to within the desired range.

A test circuit based on a rectifier circuit can provide all the necessary stresses in one step of testing without additional (voltage or current) sources. Through the control of the AC side impedance, the transformer ratio and the DC side impedance, the desired current, voltage and energy can be produced. There is a wide variety of rectifier circuits in use today, designed either for high-current or high-voltage testing. Although relatively simple to design, there is none as of yet designed for high-power applications for many reasons, be it the need, cost[2], availability of sufficient short-circuit power source, etc.

For this reason only low-voltage rectifier circuits, either for charging a reactor or a capacitor, have been used for testing of HVDC CB. For instance, a 6 kV controlled rectifier circuit connected to a short-circuit generator is used to charge a large DC reactor of up to 1 H to produce a DC current that can be used for testing the current interruption capability of HVDC CB [EHL+76].

Lately, a back-to-back 12 pulse controlled rectifier connected to the power grid was used to verify the nominal current interruption capability of an active current injection HVDC CB [She08]. First, the rectifiers are controlled to produce the DC test current (up to 4.3 kA) through a smoothing reactor at a relatively low driving voltage. In this way, an HVDC CB used as a transfer switch in a practical point-to-point project was tested before deployment. Due to the impact that this method may have on the AC supply this method cannot be used for fault current interruption tests of HVDC CBs. Similarly, a

---

[2]since the test circuit components must be rated for high current and voltage

back-to-back converter/rectifier with a large current smoothing reactor (1 H) in between has been used as a test circuit in [CIG78] and the references therein.

In another study, a fully controlled six pulse rectifier circuit made up of thyristors is used to charge a large capacitor [YD12]. A 475 mF capacitor charged to 2 kV is made to discharge through a medium sized inductor (55 mH) connected in series with the device under test (DUT) to produce very low frequency current to imitate DC current. A test current of up to 5 kA is achieved with this method. The use of a controlled rectifier indeed eliminates the impact of a test operation on the grid; however, the extremely large capacitor (475 mF) used in this method is rather difficult to obtain or impractical.

A three phase Graetz rectifier connected to a short-circuit generator is used to test HVDC CB in [STH$^+$84]. The rectifier can provide the necessary current. However, it cannot withstand the TIV produced by the HVDC CB. Thus, after charging a reactor to a required current, a crowbar switch[3] is closed to bypass the rectifier, and at the same time a back-up CB on the AC short-circuit generator side is opened. This is, in fact, a variant of a charged reactor test method. Since the output voltage of the rectifier is low, the rate of charging is slow and hence, the current through the test CB increases slowly. One of the main drawbacks of this method is the reduction of test current due to arcing in the DUT. Large reactors with high quality factors are needed.

An HVDC test station using a full-wave mercury arc rectifier connected to a high-voltage substation is reported to charge a reactor with an inductance up to 1.9 H with current in [FHKW69].

6.3.2. High-voltage charged capacitor banks

Test circuits supplied by charged capacitor banks are the most commonly used throughout the development of HVDC CBs [STK$^+$79, YTI$^+$82, CVH$^+$82, Pre82, STH$^+$84, LSY$^+$85, VL86, HJ11, GDPV14, ZWZ$^+$15, TZH$^+$, NMA$^+$19]. A charged capacitor bank is discharged through a reactor to produce short-circuit current. The reactor limits the rate-of-rise of the discharge current to within the range that the HVDC CB can handle while transferring energy from a capacitor to the DUT. A typical charged capacitor test circuit is shown in Figure 6.4 along with an example charging circuit (shown in red dashed box). It can be seen from the diagram that a test circuit based on a high-voltage charged capacitor bank is relatively simple to design – especially, it does not require a test facility having large short-circuit power. It consists of a charging circuit, a capacitor bank, MS

---

[3]Normally a crowbar circuit is used for preventing an overvoltage or surge condition of a power supply unit from damaging the circuits attached to it. However, in this sense a crowbar circuit is used for protecting the power source itself while trapping the magnetic energy in the circuit inductance (of the remaining part) from returning back to the source.

as well as safety and protection systems.

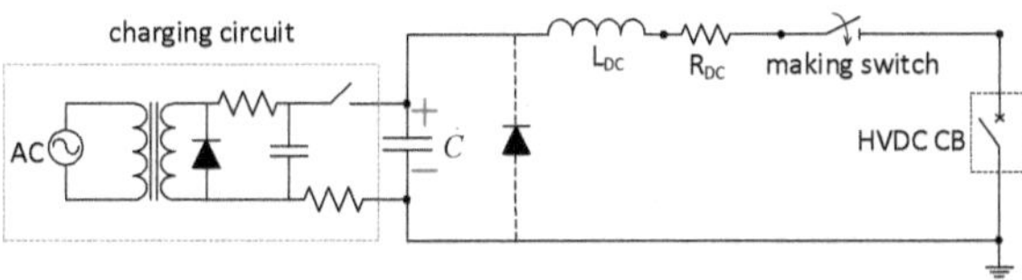

Figure 6.4.: Test circuit supplied by a charged capacitor

Because of its simplicity, a test circuit based on a charged capacitor has been (and will continue to be) extensively used for basic research and development tests of various concepts of HVDC CBs. Usually current interruption is performed around the peak value of the prospective discharge current. However, from that moment on, the capacitor bank starts to re-charge in reverse polarity and hence, retrieves part of the magnetic energy in the circuit.

In order to mitigate the retrieval of energy, a string of crowbar diodes is connected in parallel with the capacitor as shown in Figure 6.4. Before the start of the test sequence, the charge across the capacitor reverse biases the diode string. During testing, the diode string prevents the capacitor from charging in the reverse polarity after fully discharging into the reactor. This solution is implemented in [TSK+80] where a large capacitor bank (5 mF) charged to 10 kV is used as a current source through a very small inductor (0.5 mH). Alternatively, the diode string could be replaced by a triggered spark gap. However, this approach is challenging to implement especially when the triggering must be actively initiated at the peak of current from the capacitor (i.e., when the capacitor is completely discharged). Similarly, a test circuit supplied by a charged capacitor bank with a mechanical crowbar switch is used in [SKT+81] for investigation during HVDC CB development.

In some cases, a discharge frequency is chosen to accommodate the long breaker operation time. For example, a low voltage (electrolytic) capacitor bank (60 mF) charged to 2 kV was used as a current source supplying as high as 10 kA current through a reactor [Pre82]. Another very large capacitor bank (10.8 mF) was tuned to 12 Hz discharge current with a series reactor [CVH+82]. Similarly, in [VL86] a large charged capacitor bank (5 mF) is discharged through a 76 mH reactor to evaluate the performance of SF_6 based HVDC CB. A large charged capacitor (8.5 mF) is used as current source through L_{DC} (of about 200 mH) to generate a low-frequency (5 Hz) [YTI+82, STK+79].

Theoretically, with proper choice of charging voltage and circuit elements, a charged capacitor bank based test circuit ($L - C$ circuit) can be designed to replicate the fault current as well as apply adequate stress to the HVDC CB. However, practically this is almost impossible as it requires quite large capacitors that need to be charged to a high-voltage equivalent to the system voltage. On the other hand, if the capacitance of a capacitor bank within a practical range is used, the voltage across the capacitor decays rapidly making the current breaking process too easy for the CB since the difference between the TIV and the voltage across the capacitor bank (U_{CB}-U_{cap}) is very large during the current suppression interval (T_{FS}) (refer to Equation (2.7)).

Nevertheless, a test circuit using a charged capacitor bank is the most commonly used to this day due its relatively simple implementation. For example, the functional tests the 535 kV HVDC CB prototypes (used in Zhangbei project) are tested using charged capacitor banks as current sources [TZH+, ZYZ+20] – see Figure 6.5. It is reported that the hybrid HVDC CBs described in Sections 3.4.2 and 3.4.4 could clear currents up to 26 kA within 2.6 ms while generating a TIV of about 810 kV. In this case thyristor valves are used as a MS to initiate a discharge of capacitors through an L_{DC}. An additional charged capacitor bank, also using thyristors as MS, is used as a current source during re-closure testing. A re-closing functionality has been tested where the breaker is re-closed 300 ms after the inception of the first short-circuit. After re-closing, the breaker interrupted a current of 9 kA. During both interruptions no DC recovery voltage was applied.

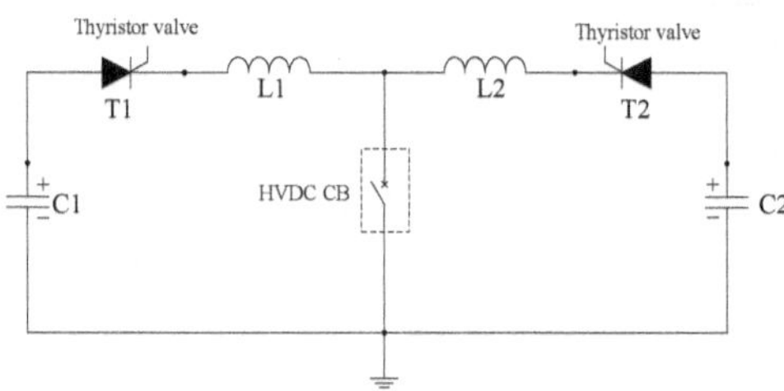

Figure 6.5.: Two charged capacitor banks used re-closing test of hybrid HVDC CB according to [TWZ+]

In [ZWZ+15], a charged capacitor test circuit is used to perform functional tests of a hybrid HVDC CB in a modular approach where sub-modules are tested individually first, followed by a test of three sub-modules connected in series. A step-by-step procedure to verify equal sharing of electrical stresses is performed before conducting an overall test of the hybrid HVDC CB described in Section 3.4. A similar test circuit and testing approach is used in [TWZ$^+$]. However, in the latter case a synthetic test approach is also employed in order to achieve a more equivalent test in terms of sufficient TIV duration as well as DC voltage stress. Hence, two high-voltage charged capacitors, one serving as current source and the other serving as voltage source, are used where two thyristor sets are used for making as well as blocking the infeed of one capacitor into the other. In such an approach, a current suppression period of over 1 ms is achieved, although this is still not sufficient.

Many variants and modifications of charged capacitor test circuits have been used to meet different purposes. For example, in order to take into account the load current just before a fault, synthetic test circuits built from the superposition of two charged capacitor-inductor circuits are used [GDPV14]. The main challenge with the use of a charged capacitor test circuit is that large capacitance values must be used in order to produce sufficient TIV and energy stresses at rated values. To tackle this challenge, a capacitor is charged to a voltage higher than the rated voltage for which the breaker is designed [GDPV14].

Another variant uses a ladder network of circuits in series (with charged capacitors) to achieve quasi-DC (trapezoidal) current for relatively long duration [Hei]. For example, three superimposed chain circuits are used to create a trapezoidal current with a plateau duration of about 6 ms. This test circuit could potentially be a good candidate for load current interruption tests. Another test circuit using multiple charged capacitor banks in parallel is described in [NMA$^+$19]. The sequential discharge of each capacitor bank through a series inductor is controlled by thyristors. This test circuit is limited to medium voltage applications.

6.3.3. High-current charged reactor

A reactor forms an integral and essential part of any test circuit for HVDC CBs. It is needed to limit the rate-of-change of the short-circuit current from a source; while it is charged, it stores magnetic energy, which the HVDC CB has to dissipate. In any test circuit a reactor of sufficient inductance is necessary to limit not only the rate-of-rise of current but also the rate-of-decay of current during the current suppression.

In most cases current sources cannot withstand the TIV stress produced by HVDC CBs. In some cases, for example, charged capacitors or AC short-circuit generators cannot supply quasi-DC current for sufficiently long duration, which is required when the HVDC CB has long breaker operation time. Moreover, if not carefully designed, these sources might draw the magnetic energy back from the circuit inductance during the current interruption process.

Therefore, in order to mitigate these challenges – the energy retraction by a current source, the excessive voltage due to the TIV on a current source and/or to produce a long duration quasi-DC current – current sources can be isolated from the circuit using a crowbar circuit in such a way that basically a charged reactor is left in a circuit. This is illustrated in Figure 6.6, which depicts a test circuit based on a charged reactor. Depending

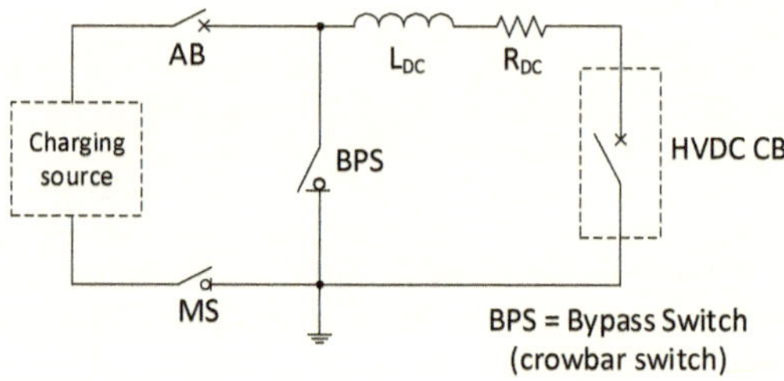

Figure 6.6.: Test circuit supplied by a charged reactor

on the type of reactor charging circuit, several realizations of charged reactor test circuits have been used for testing of HVDC CBs [HLK73, PMR+88, BMR+85, SKY14, SYBS15]. For example, a charged capacitor can be used as a current source [SKT+81].

In theory, a low voltage DC source, for example, a low voltage battery bank as in [HLK73], can also be used for charging the reactor, and the interruption process can be started once the current reaches the steady state value. Since the inductance of a reactor does not play a role in steady state conditions, the charging current magnitude is determined by the battery voltage, internal resistance and coil resistance. Once the reactor current reaches the desired level, a crowbar circuit is closed, and the battery is disconnected from the circuit by a simple contactor.

Similarly, a low-voltage rectifier could be used to charge a reactor. For example, a single phase full-wave rectifier connected to a short-circuit generator is used to charge a reactor for testing of a 500 kV, 4000 A passive oscillation HVDC CB [PMR+88]. Before opening the test breaker, the rectifier is short-circuited by an auxiliary breaker. In such a way the rectifier as well as the other components of the test circuit are protected against the 800 kV TIV produced by the DUT. Also, a rectifier (rated at 3-4 kV) is used to charge a large reactor of 0.5 H until steady state DC current is reached [STH$^+$84, BMR$^+$85]. Then before the operation of the DUT, the rectifier is decoupled from the reactor by short-circuiting its output terminals.

Other methods of charging a reactor include asymmetry transfer in four-wire three phase AC systems where the DC component of each phase is transferred to a neutral conductor as described in [SKY14, SYBS15]. To obtain the maximum asymmetry, the MSs in each phase are closed at voltage zero. By having a high-quality factor reactor in the neutral line, quasi-DC current of sufficient magnitude and duration can be achieved.

A test circuit supplied by a charged reactor can produce the pseudo-DC current required during a current interruption process. However, it lacks an intrinsic voltage source during the energy dissipation phase, as a result of which the current is interrupted faster than would be the case in service, making the test inadequate for the DUT if a proper inductance is not chosen [BS17]. For instance, if the HVDC CB has an arcing branch, the energy drawn by the arc shall not be too high to affect the current [CIG78]. A possible solution is to charge a reactor that has a large inductance [BHK74, CIG78]. This can be designed based on Equation (2.7) by setting U_{DC} to zero since there is no source voltage driving the current in this case. Referring to Equation (2.7) and assuming 1.5 p.u. TIV, three times as large a reactor as an equivalent reactor used in a test circuit supplied by a stiff DC source is needed to obtain the same stresses (TIV duration and energy) during the current suppression phase.

Moreover, the rate of reactor current decay in such a circuit depends significantly on the circuit resistance and hence requires reactors with a very high quality factor (Q). The other limitation is the charging circuit. For example, in order to use the charged reactor method it was challenging to charge a 60 mH reactor to 5 kA [GBK72]. Due to the challenges associated with the charging circuit and the fact that high-current, high-quality factor reactors are difficult to obtain, a test circuit using a charged reactor is mainly used for testing load current interruption of HVDC CBs. Also high-current reactors must be robust against inter-turn forces during test current flow since the magnetic forces between the turns of a winding increase quadratically with current.

6.3.4. AC short-circuit generators – with variable frequency

The use of AC short-circuit power sources for testing HVDC CBs is not new. It has been used extensively throughout the development of HVDC CBs, especially by operating AC short-circuit generators at low power frequency [TSK$^+$80, SKT$^+$81, TAY$^+$85]. For example, a short-circuit generator run at low speed to produce 10 Hz AC current was used to test a high-current (80-130 kA) DC CB in [TSK$^+$80]. In this case asymmetrical current was available for up to 80 ms to mimic DC current. Similarly, tests using an

AC short-circuit generator operated at a power frequency of 6 Hz together with step-up transformers were performed in [SKT+81]. Also, an AC short-circuit generator at nominal power frequency of 50 Hz has been used for testing recently developed HVDC CBs [EBH14, TKT+14, TSK+15, KHK+19].

Figure 6.7 shows a simplified schematic of a test circuit based on an AC power source along with the necessary circuit components: AC voltage source, master breaker (MB), MS, equivalent circuit impedance (R and L) and DUT – the HVDC CB. The AC short-circuit generators are represented as a simple AC voltage source behind an impedance.

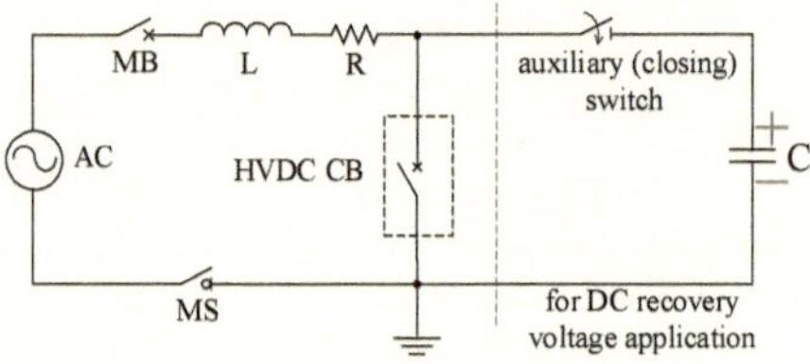

Figure 6.7.: Equivalent circuit of a test circuit using AC generator(s). ©[2017]IEEE.

6.4. Summary and Conclusion

The requirements of a test circuit are defined based on a system simulation study. A test circuit shall provide all the necessary stresses (current, energy and voltage) at the right time at each stage of the current interruption process preferably in a single test. In the absence of adequate test circuit capable of fulfilling these requirements, alternative test methods are considered: component (e.g., a current branch) testing, unit (e.g., module) testing, multi-part testing and field testing. However, a complete HVDC CB must be subjected to the worst-case situation (stresses) as possible in service.

In this chapter, various test methods and test circuits have been evaluated with respect to the critical stresses defined by the system simulation study. Ideally, a test circuit based on a DC source (such as a high-power rectifier) with adequate short-circuit power is preferred. However, such a test circuit is not available at any test facility yet, and it is unlikely that it will ever be viable from a commercial perspective. Charged capacitor based test circuits will remain essential for research and development of various concepts of HVDC CBs or components. However, a practical test circuit based on a charged capacitor cannot supply complete stresses (sufficient energy and hence, TIV duration) since the realistic energy that can be stored in a capacitor is limited. Charged reactor based test circuits can be used to verify the load current interruption performance of HVDC CBs. They cannot be used for demonstrating short-circuit current interruption performance of several kiloamperes since large reactors capable of carrying such currents are practically unavailable. Medium sized reactors indeed form an essential part of any test circuits used for HVDC CBs.

7. The Proposed Test Method – based on AC Short-circuit Generators

In this chapter the test method proposed in this **book** is discussed. After a brief introduction in Section 7.1, the method is described in Section 7.2. The detailed theoretical analysis of the proposed test method is presented in Section 7.3. In addition, a multi-part testing method proposed in case the available short-circuit power is not sufficient is discussed in Section 7.4. In Section 7.5 a simulation based comparison of the performance of various test circuits that can be used for testing HVDC CBs is presented. Finally, the conclusions drawn based on the discussion of the chapter are presented in Section 7.6.

7.1. Introduction

A test circuit based on AC short-circuit power sources discussed in the literature mainly focuses on producing a quasi-DC current for maximum duration while the magnitude of the driving voltage of the test circuit is not of prime concern. The main purpose of the tests is the proof of overall functionality of HVDC CB concepts, which are:

- achieving local current interruption in the CCB

- internal current commutation

- producing the required TIV

In those tests current interruption is deemed as successful if the HVDC CB interrupts the peak AC current well before its natural current zero as shown in the Figure 7.1. However, it can be observed from the figure that, when the AC current is at its peak, the source voltage is nearly zero and soon changes its polarity. At this stage, the short-circuit generator starts to absorb the magnetic energy in the circuit inductance instead of supplying electric energy during current suppression period unlike in an actual HVDC grid where the recovering system voltage injects energy as discussed in Chapter 4. This makes the current suppression duration shorter than in real situations especially when

the inductance in the circuit is small as can be justified in Equation (2.7). As a result the
HVDC CB is subjected to a shorter duration TIV and/or lower energy stresses, which
reduces the severity of the test. Some mitigation mechanisms such as increasing the circuit

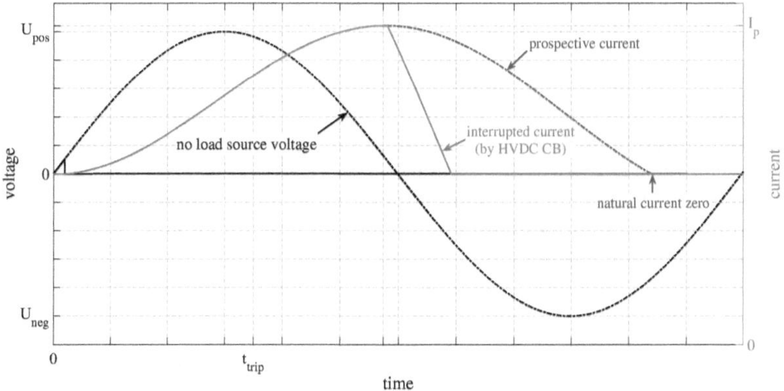

Figure 7.1.: Traditional AC short-circuit generator supplied test method for HVDC CB
testing

inductance and using a crowbar circuit as in a charged reactor test circuit discussed in
subsection 6.3.3 are introduced to reduce the impact of source voltage reduction during
this period. Also, reducing the power frequency of the short-circuit generator reduces
the energy, which is retracted by the short-circuit generator, since the voltage decreases
slower.

7.2. The Proposed Method

It is discussed in Chapter 4 that the system voltage starts to recover from the moment
an HVDC CB generates the TIV and fully recovers at the end of the current suppression
period. The recovery of the system voltage has a significant impact on the duration of a
TIV that the HVDC CB sustains and hence, on the total energy absorption. Moreover,
an HVDC CB must act fast (within 1-3 ms) to interrupt a fault current on its rising
edge well before the current reaches its steady state value [BPS18b]. The latter justifies
the fact that a steady state DC current is not a requirement in testing the fault current

interruption performance of HVDC CBs. Taking these facts into account a test circuit supplied by AC short-circuit generators operated at low power frequency, as shown in Figure 6.7, can be designed to provide adequate stresses to the HVDC CB as in service condition [BS17, BPS19].

Therefore, in this book, a test method that supplies quasi-DC source voltage instead of focusing only on quasi-DC current is proposed. In the proposed method the desired test current is produced and interrupted while the source voltage is near its crest value. This is illustrated in Figure 7.2, which juxtaposes experimental (left) and simulation (right) results of a test circuit by AC short-circuit generators running at 16 2/3 Hz. The left y-axes in both graphs show voltage measurements while the current measurements are on the right y-axes of each graph. The equivalent inductance of the test setup, including the sub-transient reactance of the short-circuit generators and the leakage reactance of the step-up transformers, is circa 25 mH.

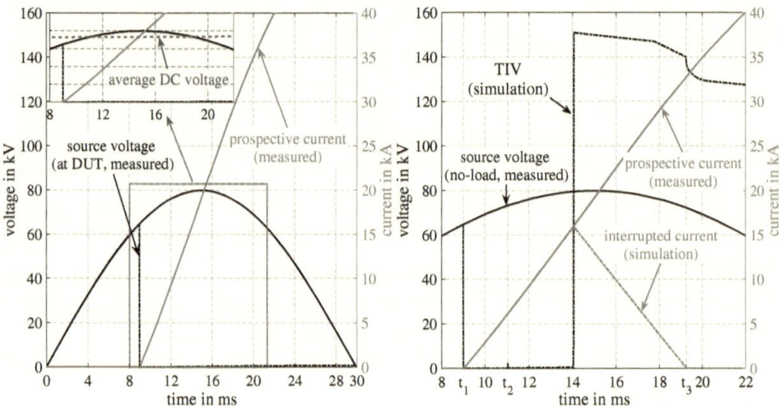

Figure 7.2.: Experimental (left) versus simulation (right) results. The experimental result is obtained by running four short-circuit generators at 16 2/3 Hz in combination with six step-up transformers. Interruption process by a model of HVDC CB inserted in a simulation test circuit is superimposed (right graph). At t_1 a short-circuit is applied (making), at t_2 the trip command to the breaker, and t_3 marks the end of current suppression.

The graph on the right side illustrates the simulation result obtained with the same circuit setting as the experimental setup (only the relevant part is shown). The four short-circuit generators (each with 2250 MVA at 50 Hz) and the six step-up transformers used in the experiment are represented by an equivalent AC source behind an impedance. Moreover, a model of a hybrid HVDC CB is embedded into the simulation setup in order to get insight into the expected electrical stresses on a real HVDC CB during the current interruption process in this test circuit. Assuming a fault current neutralization time of about 5 ms (of which 2 ms is a relay time) 16 kA current is interrupted. Moreover, short-circuit current rising at a rate of 3.2 kA/ms is achieved.

In addition to the short-circuit power, the proposed method depends on the precise control of the following parameters of an AC test circuit,

- Power frequency (f) – rotor speed (N, rpm) and number of magnetic poles of stator winding (P) $[f = \frac{PN}{120}]$

- Making angle (θ)– a point on voltage wave (in reference to voltage zero), at which the short-circuit is applied (controlled using MSs synchronized with rotor angle)

- Voltage amplitude $(\widehat{U})$– Combination of generator excitation and transformer ratio

- Circuit impedance (Z) – equivalent (reactive) impedance[1] including generator transient reactance, adjustable current limiting reactance and transformer leakage reactance, connection (rail) impedance, etc.

At most test facilities, the above parameters are known and can be controlled (varied within sufficient margins) not just for the purpose of testing HVDC CBs but as readily implemented features for flexibility of testing various types of AC equipment.

In order to increase the duration of the desired source voltage, tests are performed at reduced power frequency, for example, at 16 2/3 Hz. At this frequency, for instance, the source voltage remains above 80% of its peak value within ±6 ms from the crest point, providing a window of 12 ms for testing, see the zoomed section in the left hand side graph of Figure 7.2. Depending on the actual speed of operation of the HVDC CB, this time window can be adjusted; expanded or narrowed around the crest point. Thus, if a short-circuit can be applied at a precise point on the voltage wave (i.e., at a precise making angle), and with proper selection of the circuit inductance, a short-circuit current with a desired rate-of-rise can be produced as shown in the figure. After the initiation of the short-circuit (equivalent to relay time), the HVDC CB is tripped to break the short-circuit current while it is on the rise, similar to the expected operation in a MTDC grid. Hence, it can be seen that the complete current interruption process occurs while

---

[1]assuming negligible resistance

the source voltage is sufficiently high – within a margin that can be determined based on test parameters. In such a way, sufficient stresses in terms of current, voltage and energy are ensured.

7.3. Theoretical Analysis of the Proposed Test Method

In [BSTI16] it is shown that a test circuit based on AC short-circuit generators operated at low power frequency can be used for testing HVDC CBs. In this section the mathematical analysis of such a test method is discussed as presented in [BPS19].

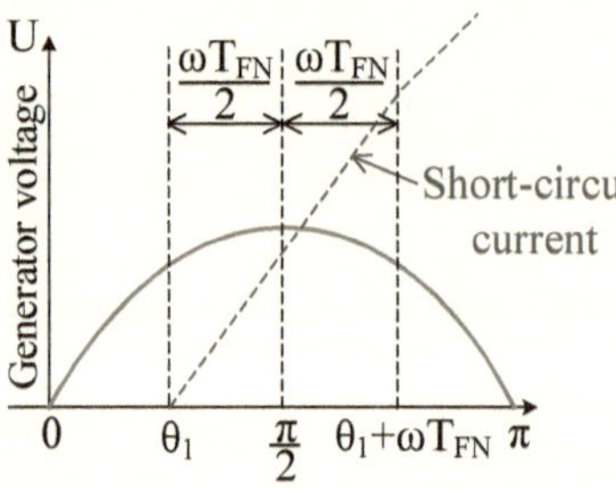

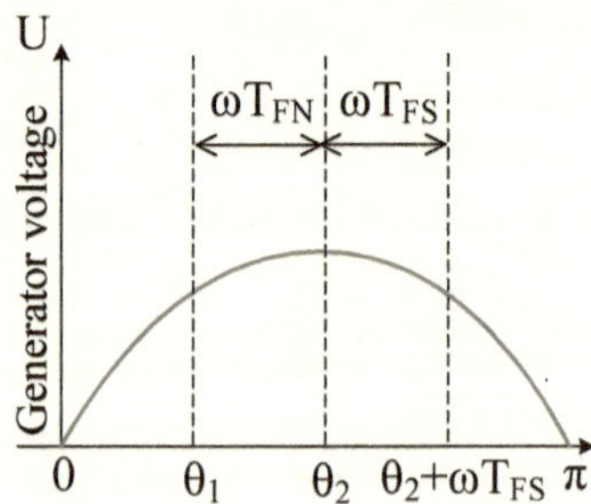

(a) Making angle for maximum average rate-of-rise of current

(b) Optimized making angle for rate-of-rise of current and energy absorption

Figure 7.3.: Illustration of a HVDC CB test method using AC short-circuit generator: Optimization of the making angle ©[2019]IEEE.

Referring to Figures 6.7 and 7.3, the governing equation during the fault neutralization time (T_{FN}) is

$$\widehat{U} \sin(\omega t + \theta_1) - Ri(t) - L\frac{\mathrm{d}i}{\mathrm{d}t} = 0 \tag{7.1}$$

where $\omega = 2\pi f$ and f is the AC generator power frequency, θ_1 is the making angle, R is the parasitic resistance in the circuit and L is the equivalent inductance in the circuit. The maximum average $\mathrm{d}i/\mathrm{d}t$ over T_{FN} is obtained when $\theta_1 = \pi/2 - \omega T_{\mathrm{FN}}/2$, see Fig.7.3a.

The solution of Equation (7.1) is,

$$i(t) = \widehat{I}\left[\sin(\omega t + \theta_1 - \phi) - \sin(\theta_1 - \phi)\exp(-t/\tau)\right] + I_0 \exp(-t/\tau) \tag{7.2}$$

where $\widehat{I} = \dfrac{\widehat{U}}{\sqrt{R^2+(\omega L)^2}}$ is the peak value of symmetric current, $\phi = \tan^{-1}(\omega L/R)$ is the phase angle, $\tau = L/R$ is the circuit time constant and I_0 is the value of the initial current flowing at the moment when the short-circuit is applied.

Assuming a negligible voltage rise time of the HVDC CB, the peak value of the interrupted current is I_p obtained from $i(T_{FN})$ in Equation (7.2). At the moment the HVDC CB starts to generate the TIV, it starts to absorb energy. This marks the beginning of the fault current suppression period (T_{FS}). The circuit equation during this period becomes:

$$\widehat{U}\sin(\omega t + \theta_2) - Ri(t) - L\frac{di}{dt} - U_{CB} = 0 \tag{7.3}$$

where U_{CB} is the TIV of the HVDC CB. Since the same AC source is supplying power during both T_{FN} and T_{FS} periods, θ_2 equals $\omega T_{FN} + \theta_1$, see Figure 7.3b, where the time reference is changed to the beginning of T_{FS} for mathematical convenience.

The EAB of the HVDC CB is assumed to have a number of parallel MOSA columns to absorb the energy stress with each column operating in a highly non-linear region of its U–I characteristics. Thus, Equation (7.3) can be solved by assuming a constant TIV during T_{FS} and the solution is expressed as:

$$i(t) = \widehat{I}\left[\sin(\omega t + \theta_2 - \phi) - \sin(\theta_2 - \phi)\exp(-t/\tau)\right] + I_p \exp(-t/\tau) - \frac{U_{CB}}{R}(1 - \exp(-t/\tau)) \tag{7.4}$$

From the solution in Equation (7.4), the duration of T_{FS} can be determined numerically by setting $i(t) = 0$ and solving for t. The energy that the HVDC CB absorbs during T_{FS} can then be estimated analytically as follows:

$$\begin{aligned}
E &= \int_0^{T_{FS}} U_{CB} i(t)dt \\
&= \frac{1}{2}LI_p^2 + \int_0^{T_{FS}} \widehat{U}\sin(\omega t + \theta_2)\, i(t)dt
\end{aligned} \tag{7.5}$$

Where $i(t)$ is given by Equation (7.4). The energy calculated in Equation (7.5) is contributed by two sources: the energy stored in the circuit inductance at the beginning of T_{FS} ($\frac{1}{2}LI_p^2$), and the energy being supplied by the AC short-circuit power source during

T_{FS}. The latter corresponds to the energy supplied by the HVDC grid during the same period as discussed in Section 5.1, which in most of the cases is larger than the former.

Similar to the maximally achievable di/dt during T_{FN}, the maximum energy that can be supplied by the source is achieved when $\theta_2 = \pi/2 - \omega T_{FS}/2$. A further increase in supplied energy can be realized by prolonging the availability of the AC source voltage by lowering the power frequency. Thus, for a given T_{FN}, a making angle of $\theta_1 < \pi/2 - \omega T_{FN}$ is a good compromise between di/dt as well as energy, see Figure 7.3b. This entails that the power frequency of a test circuit must be chosen such that $T_{FN} > 1/(4f)$. The AC source voltage is not directly applied to the HVDC CB during current interruption. However, it follows from the above analysis that it is a crucial test circuit design parameter during both periods (T_{FN} and T_{FS}), and its magnitude must be carefully determined and distributed over the two periods.

Although the test method based on AC short-circuit generators is the most suitable, and the required test circuit components are readily available in most existing AC short-circuit test laboratories, it is not complete in itself. Additional features such as the ability to apply DC recovery voltage after current suppression are required to enable comprehensive testing of the current interruption performance of HVDC CBs. The practical implementation of a complete test circuit based on AC short-circuit generator is discussed in Chapter 8 along with many additional features.

7.4. Proposed Multi-part Testing Method

In Chapter 2, it was shown (analytically) that the stresses of HVDC CBs are interdependent in that one cannot be applied independent of the other. For instance, complete and rated stresses are ensured only if the rated energy is applied at the rated short-circuit current. Rated TIV parameters, including rate-of-rise, peak value and duration, are automatically fulfilled under this condition. However, in some cases, especially when testing extra-high voltage HVDC CBs, it may not be possible to apply full rated energy due to either a limitation of a test laboratory or because for some practical reasons the test breaker is supplied with reduced energy rating. Then, a rated energy absorption test cannot be performed, and this affects the other stresses, in particular the TIV duration. In such a case, an alternative approach that can replicate the TIV magnitude and duration, as would be during rated current interruption test at rated energy (while actually a test is performed at reduced energy), is sought.

To address this, an equivalent two-step test procedure is proposed in this book. The first test is performed to obtain the initial transients as during the rated energy[2] test by

---

[2]the energy that would be absorbed when a source voltage equivalent to the rated system voltage is

performing a current interruption test at rated short-circuit while the actual energy is lower than the rated energy. In this test the duration of the TIV depends on the specified energy absorption of the test breaker. It can be too short compared to service operation. In the next test, a TIV duration, which complies to the service operation condition, is targeted. Thus, the energy absorption may not be at rated value but the TIV duration can be of the realistic duration. This is achieved by performing a test at lower current than the rated peak interruption current. Because the energy has a square relation with the peak value of the breaking current, significant reduction in energy while ensuring nearly rated duration of TIV can be achieved. This is illustrated in Figure 7.4.

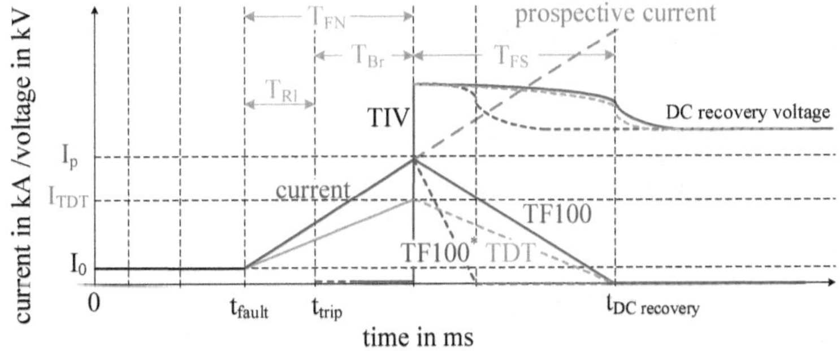

Figure 7.4.: Illustration of multi-part testing method – ensuring complete stresses in two steps

TF100, shown by the solid blue lines (for both voltage and current) in Figure 7.4, represents the rated short-circuit current breaking (I_p) at rated energy and hence, exhibits realistic $\mathrm{d}u/\mathrm{d}t$, peak value and duration of TIV. However, if this test cannot be performed in one step, it can be performed in two steps – first TF100* (shown by the dashed blue lines) and then perform TDT (shown by the green dashed lines). During TF100* test, current I_p is interrupted as in the TF100 test. This test represents rated maximum current interruption capability at maximum $\mathrm{d}u/\mathrm{d}t$ and peak value of TIV but for short duration because of the reduced energy. The duration of the TIV in this test is determined based on the specified energy absorption ($E_{\text{specified}}$) of the DUT. On the other hand, during

---

used in the test

TDT test, current I_{TDT} is interrupted. The I_{TDT} and other test circuit parameters such as the source voltage magnitude $\widehat{U}$ and circuit inductance L are determined from $E_{\text{specified}}$ and the rated T_{FS} of the DUT according to Equation (7.6).

$$\frac{1}{2}L_X I_X^2 + \int_0^{T_X} \widehat{U}_X \sin\left(\omega t + \theta_2\right) i(t) dt \leq E_{\text{specified}} \tag{7.6}$$

where subscript X stands for either TF100* or TDT. During TF100* test, I_X equals I_{p} whereas during TDT test, T_X equals rated T_{FS}. The remaining parameters in each test are numerically determined based on the discussion in section 7.3.

Examples of actual test results illustrating this approach are discussed in Chapter 8.

7.5. Comparison of Test Circuits

In order to better understand the differences in performance as well as to get an insight of practical feasibility of the test circuits described above, a simulation comparison is performed in this section. The simulation parameters for each test circuit are chosen considering the actual values at a test laboratory. A test circuit supplied by an ideal DC source is included to provide a reference for comparison. The simulation model of an 80 kV HVDC CB is used for short-circuit current interruption assuming a breaker operation time of 2 ms.

Figure 7.5 shows simulation results when 9 kA current is interrupted in each test circuit. In order to illustrate the differences, an equivalent circuit inductance of 20 mH is used in each test circuit. In addition, the peak values of AC source voltages as well as the charging voltage of the charged capacitor test circuit is set to 80 kV. In the latter case a capacitor bank of 300 μF is charged. Simulations are performed disregarding protection relay time in order to focus only on the differences between the basic electrical stresses supplied by the different test circuits. Thus, considering the plots in Figure 7.5a there is a slight difference in the rate-of-rise of short-circuit current during the fault neutralization period especially for the charged reactor and charged capacitor test circuits. Hence, each of the four circuits can generate similar conditions during fault neutralization.

The main difference among the various test circuits is observed during the fault current suppression phase, which results in a significant difference in the duration of fault current suppression period. As a result, the duration of the TIV, one of the critical stresses of the HVDC CB, varies depending on the test circuit – see Figure 7.5b. This difference is attributed to the considerable differences in the amplitude of the source voltages of the test circuits during the energy absorption phase as shown in the Figure 7.5c. It can be seen from this graph that the charged reactor test circuit does not have an intrinsic source supplying electrical energy whereas in the case of charged capacitor test circuit, the capacitor has lost a significant portion of its stored charge before the current suppression starts. For the AC supplied test circuits, the making angles are chosen such that the peak values of the source voltages appear during the current suppression period. In addition, the making angles are chosen considering the desired rate-of-rise (compared to ideal DC source) during the fault neutralization period.

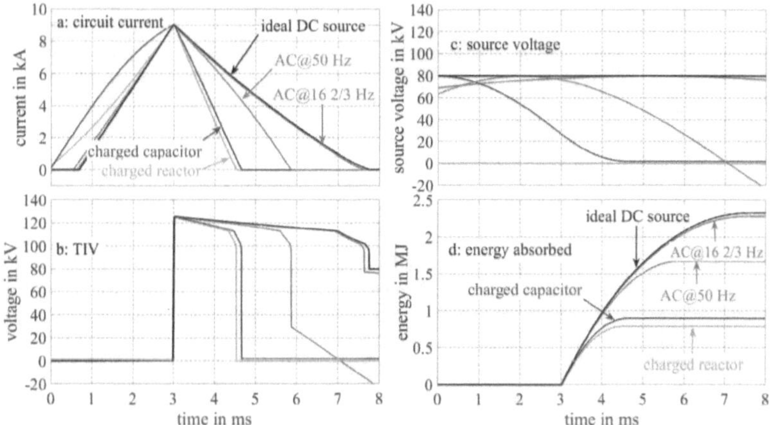

Figure 7.5.: Comparison of simulation results of various test circuits. ©[2017]IEEE.

The difference between the test circuits is also clearly reflected in the amount of energy absorbed by the HVDC CB as shown in Figure 7.5d. Since the energy initially stored in the circuit inductance ($LI_\mathrm{p}^2/2$) at the beginning of fault current suppression is essentially identical for all the test circuits, the additional energy dissipated in the CB is contributed by the voltage sources of the test circuits as described by Equations (2.11) and (5.1).

Theoretically, any of the test circuit parameters can be designed to fulfill current, voltage and energy stresses of an HVDC CB. However, when considering the practical feasibility, with respect to high-current interruption and large energy absorption, the realization of all the test circuits inevitably becomes challenging especially as the rating of the HVDC CB increases and/or when its breaker operation time increases. Even for the lower range of high-voltage modules, the test circuits supplied by the charged reactor and charged capacitor are challenging to realize especially for sufficient duration of TIV as well as adequate energy stresses.

The main conclusion from the simulation comparison of the test circuits shown in Figure 7.5 is that a test circuit using AC short-circuit generators operated at low frequency can be adequate for testing HVDC CBs. Compared to charged reactor and charged capacitor test circuits, a 50 Hz AC generator based test circuit has better performance because of the additional flexibility due to the controllable making angle. However, when evaluated with respect to an ideal DC and 16.7 Hz AC source test circuits, the 50 Hz AC source lacks sufficiently long duration source voltage to provide sufficient energy stress to the DUT. The main features as well as limitations associated with each of the test circuits is summarized in Table 7.1. Nevertheless, as illustrated in the preceding section, care must be taken when using AC sources for HVDC CB testing. The voltage magnitude, and hence the making angle, must be chosen in such a way that the crest of the source voltage appears during the energy absorption phase (current suppression time) of the interruption process. In this way sufficient energy contribution from the source, and as a result, adequate stress on the energy absorption component of the breaker is guaranteed.

For all the test circuits except the ideal DC source, the post current suppression DC recovery voltage stress is missing. The recovery voltage is an essential part of HVDC CB stresses and must be applied. The techniques to address this and many other practical challenges associated with the implementation of a test circuit based on AC short-circuit generators are discussed in Chapter 8.

7.6. Summary and Conclusion

A realistic test method which utilizes readily available test circuit components at most short-circuit test facilities is proposed in this chapter. The test method is based on the use of AC short-circuit generators operated at low power frequencies, in this book at 16 2/3 Hz. The test method maintains the supply of quasi-DC voltage while providing short-circuit current with sufficient rate-of-rise and magnitude. Since the power generated by short-circuit generators is based on mechanical energy stored in its rotating mass, a large amount of electrical energy can be extracted. The test circuit has been designed and demonstrated at various the maximum voltage and current ratings. It is verified that this test method can supply the required current, energy and voltage stresses during the current interruption process. A multi-part testing alternative is proposed to demonstrate capability of an HVDC CB to withstand and sustain the TIV even when, due to practical limitations, a test breaker is not equipped with the full energy absorption module that is producing such a TIV for a rated duration.

Table 7.1.: Summary of comparison between test circuits

Test Circuit	Features	Limitations	Remarks
Rectifier 6.3.1	• Can reproduce full stress – current, voltage and energy stresses • Can supply DC recovery voltage after current suppression	• Requires high-power rectifier converter • Not readily available for HVDC CB testing • Requires high short-circuit power	• Limited to low-to-medium voltage DC CB testing • Suitable for long duration DC current withstand test
Charged Capacitor 6.3.2	• Readily available at test facilities • Mostly used to demonstrate DC current interruption (proof-of-concept) • Can realize very high di/dt test current • Simple design, implementation and execution	• Very low energy leads to very short TIV duration • Requires large capacitance for HVDC CBs testing • Large volume, high cost, and safety	• Does not require short-circuit power source • Can be combined with other test circuits
Charged Reactor 6.3.3	• Can produce long duration quasi-DC current	• Accurate switching of charging circuit • High quality factor and robust design of reactor for high-current • Powerful source required for charging to high currents • Limited energy and short TIV duration	• Low-to-medium current and energy tests • Suitable for load current interruption test • Lacks intrinsic voltage source
AC short-circuit generator 6.3.4	• Readily available at test facilities • High-current at various di/dt • Can be designed to provide full-stress	• TIV stress may exceed the insulation coordination of the test installation • Maximum di/dt is limited (minimum inductance due to generator and transformer reactance) • Output voltage and power reduces with frequency	• Full current, high energy testing • Adjustable power frequency is essential

8. Design and Implementation of a Test Circuit Based on AC Short-Circuit Generators

In this chapter a practical implementation of a complete test circuit based on AC short-circuit generators is elucidated following the method proposed in Chapter 7. In addition, testing of HVDC CBs using AC short-circuit generators poses new challenges; namely, the protection of both the DUT as well as the test circuit components when the HVDC CB fails to interrupt, and the application of DC recovery voltage after current suppression. Pragmatic methods to overcome these challenges are developed and demonstrated in the test laboratory. Moreover, tests validating the proposed method and circuit have been conducted on three different prototypes of HVDC CBs with different level of modularity and with ratings up to 350 kV, 20 kA. The adequacy of the test circuit is evaluated based on the actual test results. Finally, some imminent challenges of testing EHVDC CBs are discussed based on the actual test results. Particularly, transformer saturation due to the HVDC CB's TIV is one of the pre-eminent challenges of using AC installation for HVDC CB testing.

Some of the contents of the chapter are based on publications [BPS19, BPS+18a]. The chapter is organized as follows. After brief introduction in Section 8.1, a complete test circuit based on AC short-circuit generators along with all the necessary additional features and circuitry is presented in Section 8.2. The four fundamental parts of the test circuit are discussed in detail together with the practical implementations and actual demonstrations of operations of each part. In Section 8.3, the performance demonstration of the test circuit is presented along with test results of the prototype HVDC CBs. Section 8.4 provides a discussion on the challenges of testing multi-module HVDC CBs and the inherent limitations of the test circuit and the test method. Finally, the conclusions based on the results of the chapter are presented in Section 8.5.

8.1. Introduction

Except the high-power rectifier test circuit, none of the test circuits discussed in Chapter 6 can supply the complete stresses to an HVDC CB as described in subsection 5.3. In Chapter 6 several test circuits are evaluated and compared on the basis of whether a given test circuit can supply the necessary stresses (current, voltage and energy) during the current breaking process. In all cases, the rated energy absorption requirement resulting from the required TIV duration and the lack of DC recovery voltage are the main limitations.

Many different approaches have been considered in order to address some limitations of test circuits. For example, trapped charges on the resonant capacitors of the mechanical HVDC CBs have been used as a source of DC recovery voltage after current suppression [STK$^+$79, Pre82, VL86]. However, this method depends on the actual design of the HVDC CB and cannot be applied to all technologies. Moreover, it is not the best test practice to utilize self-imposed stresses of the internal components to replace system imposed stresses. In general, an adequate test circuit capable of replicating the complete stresses in a test, independent of the design and the technology of the HVDC CB, is preferable. With this regard, the main challenge today is to design and implement not only a technically feasible but also an economically sound test circuit that can adequately and safely supply the complete stresses as in service.

8.2. Description and Implementation of the Test Circuit

The test method is based on the use of AC short-circuit generators operated at low power frequency as depicted in Figure 8.1. Considering a half-cycle of a sinusoidal AC voltage at frequency of 16 2/3 Hz (see dashed red trace in the figure), a quasi-DC voltage, which remains within 75% of the crest value for about 14 ms, can be used for testing. Thus, by applying a short-circuit at a precise point on the voltage wave and with proper selection of circuit impedance, a short-circuit current with adequate rate-of-rise can be produced. Coupled with the fact that the HVDC CBs operate quite rapidly (within several milliseconds), the desired test current can be produced and interrupted while the source voltage is near its crest value[1]. The critical design parameters of a test circuit, enabling the application of the necessary stresses: current, energy and voltage during the current breaking process, are discussed in Section 7.3.

Of all the critical stages of the fault current breaking process described in Section 5.3,

---

[1]The entire current breaking process occurs while the source voltage is sufficiently high – seen as a DC source by an HVDC CB.

only stage 4 (DC system recovery voltage) cannot be applied by the AC short-circuit generator. This has to be applied in a "synthetic" way, i.e., from a separate DC voltage source such as a pre-charged capacitor bank.

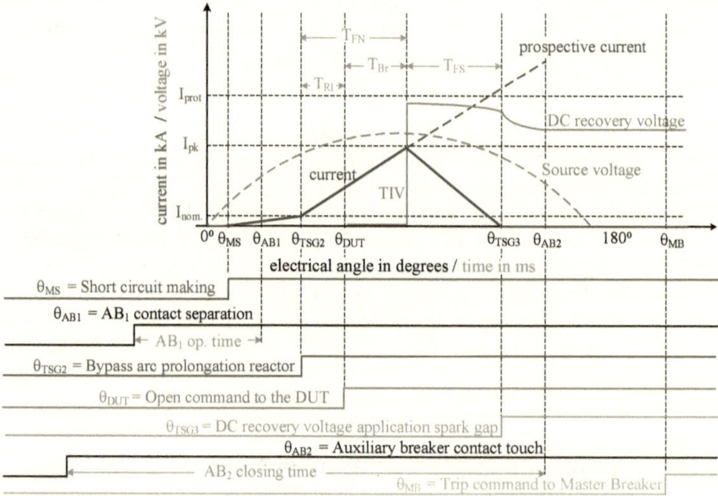

Figure 8.1.: Generic current and voltage waveforms with detailed timing indications

A schematic of a complete test circuit based on AC short-circuit generators is shown in Figure 8.2. The test circuit is composed of four parts: a short-circuit power source, an overcurrent protection circuit, an arcing time prolongation circuit and a DC recovery voltage source, each of which are indicated in separate dashed boxes. The sequence of operation of the switching components in each part is illustrated along with generic current and voltage waveforms in Figure 8.1. The detailed operation and the purpose of each part is described in next sections.

8.2.1. Short-circuit power source

This is the main source of the short-circuit current, voltage and energy during the current breaking process. It consists of AC short-circuit generator(s) and short-circuit power

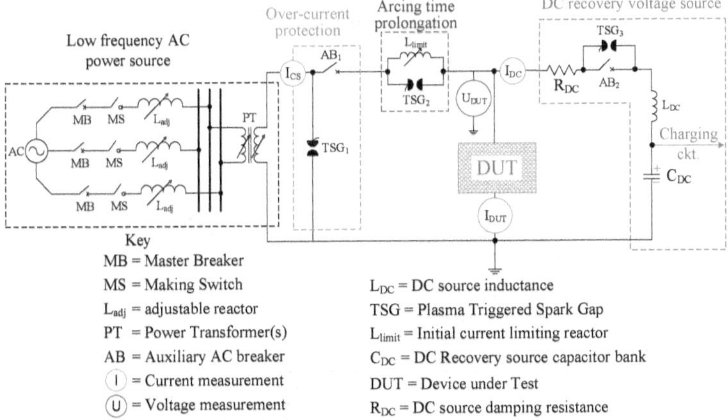

Figure 8.2.: Schematic of the complete test circuit for testing short-circuit current interruption performance of HVDC CB

transformer(s) both capable of operating at low power frequency. A short-circuit generator has a master breaker (MB) and a making switch (MS) in each phase as shown in Figure 8.2. For testing HVDC CBs, a single phase circuit is used[2] and only one (major) loop[3] of AC current is needed. The MSs are used to precisely control the making angle (θ_{MS}) whereas the MBs clear the short-circuit current (if it exists) at the first power frequency current zero (a few milliseconds after θ_{MB}) as shown in Figure 8.1. In each phase, an adjustable series reactor (L_{adj} in Figure 8.2) with multiple taps is available. These reactors, together with the generator(s) sub-transient reactance and the transformer(s) leakage reactance, allow adjustable short-circuit impedance to obtain the short-circuit currents with a desired rate-of-rise and store sufficient energy to be dissipated in the HVDC CB.

When operating short-circuit generators at low power frequency, the generated voltage and hence, the available power are reduced proportionally with frequency. Therefore,

---

[2]actually connected between two phases as shown, which reduce to a single phase with the resulting line-to-line voltage

[3]short-circuit current from initiation until the first natural current zero or between two successive current zeros

multiple series-connected power transformers are needed to step-up the test voltage to a desired level, and several short-circuit generators are connected in parallel to compensate for the reduced short-circuit power[4]. A simplified electrical diagram of a typical connection of multiple generators and transformers is shown in Figure 8.3 [BSTI16]. The short-circuit power transformers can be connected in parallel to adjust (reduce) the total circuit impedance; thus, providing additional flexibility besides adjustable reactors on the low voltage (primary) side. In Figure 8.3 two groups of short-circuit power transformers are connected in parallel where each group consists of five series connected transformers.

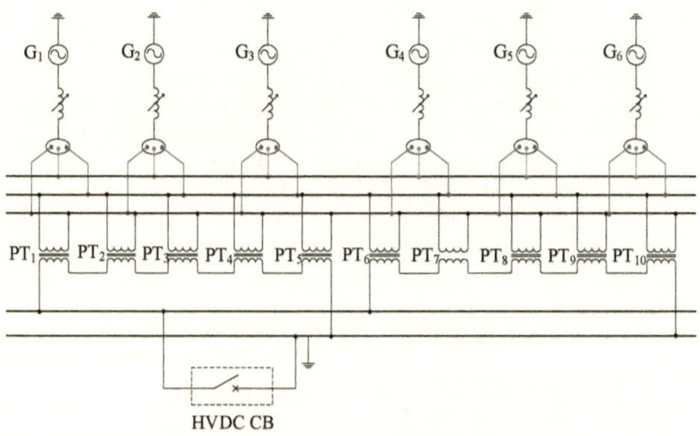

Figure 8.3.: An example of connecting six short-circuit generators (G_1-G_6) and ten power transformers (PT_1-PT_{10}) for testing HVDC CB. ©[2017]IEEE.

8.2.2. Overcurrent protection

The drawback of using AC power sources operated at low power frequency is that there exists a possibility where significant prospective current flows in the circuit (through DUT) for tens of milliseconds. Thus, if a DUT does not operate for any reason or fails to clear, it will be subjected to the full prospective asymmetric (major loop) current from the power source (until the MB clears). Even if only a single major loop current flows during a test, it could result in a damage to the DUT and/or to the components of the test installation.

---

[4]This also compensates for the reduced current at the transformer's high-voltage terminal (DUT side).

To mitigate this risk, an overcurrent protection circuit has been implemented – see
the dashed box labeled as overcurrent protection in Figure 8.2. It consists of a plasma
triggered spark gap TSG_1 and an auxiliary HVAC SF_6 CB AB_1. The TSG_1 is controlled
by a real-time current level detector, which sends a triggering signal if a pre-set threshold
value is exceeded. This is demonstrated in the test laboratory as shown in Figure 8.4. In
the results shown, the generator produces a current with a prospective peak of 33.5 kA
and loop duration of 45 ms. The detection threshold of the level detector is set to 20 kA
in this case.

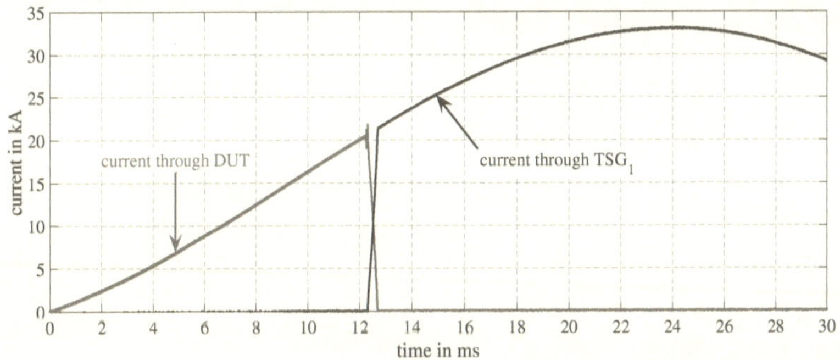

Figure 8.4.: Demonstration of successful overcurrent protection by triggered bypass spark
gap TSG_1 connected in parallel with the DUT. ©[2019]IEEE.

Normally the DUT would have cleared the short-circuit current at 16 kA, but in this
case the DUT was intentionally kept in a closed position to demonstrate the overcurrent
protection circuit. Thus, the TSG_1 is triggered when the current threshold of 20 kA is
detected, at which point the current begins to commutate from the DUT to the TSG_1
(commutates in about 400 μs), preventing overcurrent which might have otherwise resulted
in a possible damage to the DUT and/or the test installation – for example, damage to
AB_1 due to long $(40 - 50$ ms) arcing.

During the actual test, AB_1 is also tripped well before the trip command is sent to
the DUT because it needs to provide galvanic isolation between the power source and
the DUT after the current is suppressed. If galvanic isolation is not provided by AB_1,
the AC source and the DC recovery voltage source can interact, thus making it difficult

to maintain a DC voltage supply to the DUT after current suppression[5]. Therefore, it is necessary to isolate the DUT from the short-circuit power source immediately after current suppression. The additional advantage of AB_1 is that its arc voltage[6] enhances the commutation of current from the DUT to the TSG_1 when the latter is triggered due to overcurrent detection. Otherwise, depending on the loop inductance between the DUT and TSG_1, current commutation might not happen at all, or only partial commutation occurs. This is illustrated in the next subsection.

Moreover, the gap length of the TSG_1 must be set to withstand the TIV under successful current breaking condition. It can also be set for passive overvoltage protection while keeping sufficient margin with respect to the expected peak of the TIV. This protects the DUT and the test circuit components against overvoltages above the TIV peak in case they occur for any unforeseen reason. However, as the minimum required gap length of the TSG_1 increases due to the increase in the TIV of the DUT, triggering of the spark gap becomes challenging. Further discussion about this is found in the appendix A.6.1.

8.2.3. Auxiliary breaker arcing time prolongation

For AB_1 to provide galvanic isolation, it must have gained sufficient dielectric strength already by the time the DUT suppresses the short-circuit current. This is because AB_1 is subjected to a voltage difference between the AC power source and the applied DC recovery voltage[7].

Any HVAC CB has a minimum arcing duration before it can reliably interrupt and isolate. For most HVDC CBs, the total current flow duration – interruption time ($T_{FN}+T_{FS}$) – is shorter than the minimum arcing duration required by the HVAC CB. Therefore, arcing in the HVAC CB must commence well before the application of short-circuit current to the DUT. This requires artificial prolongation of current flow duration before the start of the operation of the DUT. In most cases, this can be achieved by utilizing the extra volt-second area just before the actual test window as shown in Figure 8.1 – between θ_{MS} and θ_{TSG_2}. In order to limit the magnitude of current flowing during arcing time prolongation, an additional reactor is inserted (see the dashed box labeled as arcing time prolongation in Figure 8.2) to let AB_1 reach its minimum arcing time at low current level. This current level can be in the order of magnitude of normal load current of the

---

[5]The interaction could also cause damage to the DC recovery voltage source installation, and vice versa damage to the AC transformers could be caused by the DC source installation.

[6]The arc voltage of AB_1 does not affect the test current and voltage. The arc voltage of AB_1 is in the order of a few hundred volts. This can be compensated for by slightly increasing the source voltage.

[7]This is critical because the differential voltage increases soon after current suppression due to the polarity reversal of the AC source voltage.

HVDC CB. Then, this reactor is bypassed by a triggered spark gap TSG_2 and the actual short-circuit current with the desired rate-of-rise is produced.

This is demonstrated in the test laboratory using a simplified test circuit with a schematic shown in Figure 8.5. For this demonstration a DUT is not necessary. However, the actual current path through the DUT must be represented by stray inductance (L_{stray}). The latter is critical for overcurrent protection as discussed below.

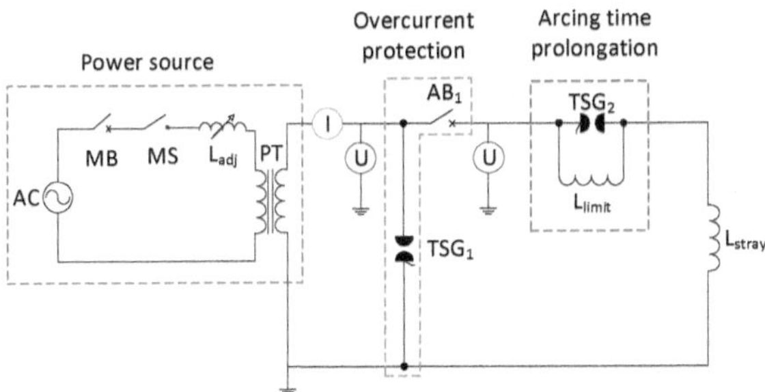

Figure 8.5.: Schematic of a part of the test circuit used for demonstration of arcing time prolongation. ©[2019]IEEE.

The test is performed at a power frequency of 16 2/3 Hz and the test results are depicted in Figure 8.6. In this demonstration, the MSs are closed at 2 ms (at 12° electrical) from the source voltage zero. A current limiting reactor L_{limit} of 81 mH is used in the arcing time prolongation circuit. Current flows through this reactor for about 7.5 ms, after which the TSG_2 is triggered to provide a bypass path for the short-circuit current. When the spark gap is triggered, L_{limit} remains charged and maintains its current, which decays by the inherent resistance of the reactor and the arc in TSG_2. AB_1 is pre-tripped so that its contacts separate when a few tens of Amperes of current starts to flow – when current exceeds the chopping level of the AB_1. L_{limit}[8] is selected so that the current chopping level of AB_1 is exceeded within <1 ms after making. The triggering of the TSG_2 initiates the actual short-circuit current, which rises at the required di/dt, in this particular case at about 2.3 kA/ms. In such a way additional arcing time of about 7.5 ms is gained for AB_1.

---

[8]L_{limit} is selected based on the required arcing prolongation time, the initial current at AB_1 contact separation and load current through the DUT just at the start of short-circuit current.

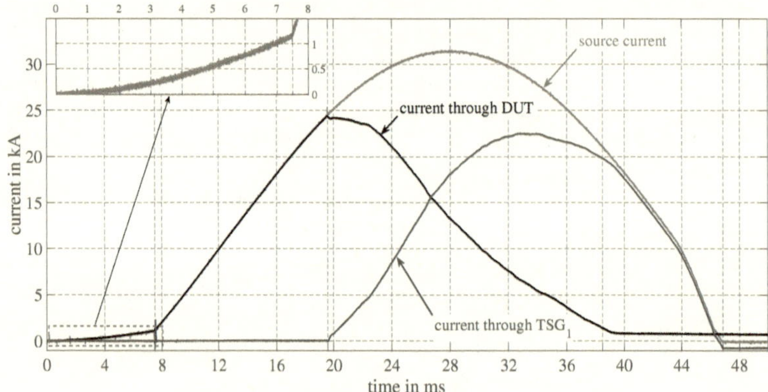

Figure 8.6.: Experimental demonstration of arcing time prolongation in the auxiliary breaker combined with verification of overcurrent protection

In this experiment, overcurrent protection is also demonstrated in combination with arcing time prolongation. The experiment is set up with a thought that the DUT will start to suppress the short-circuit current when it reaches 16 kA – assuming a fault current neutralization time of 7 ms. In order to demonstrate the overcurrent protection, the threshold value is set to 25 kA. Thus, when the current reaches 24.5 kA, a real-time current level detector sends a trigger signal to the TSG_1. Then current starts to commutate from the DUT path (see the black trace) to the spark gap bypass path (see the blue trace).

Compared to Figure 8.4 the current commutation time during overcurrent protection shown in Figure 8.6 is significantly longer. This is due to the difference in the test setups, which resulted in a larger stray inductance L_{stray} in the loop between the DUT and the TSG_1 – see Figure 8.5, in the latter case. In addition, in the latter case the AB_1 is not tripped and hence, there is no arc voltage enhancing the current commutation. Nevertheless, experimental investigation has shown that it is not only the arc voltage of AB_1 that determines current commutation, but also the stray inductance in the DUT path.

The impact of the stray inductance can also be seen towards the end of the current commutation in Figure 8.6. Even if the entire current through TSG_2 is commutated (at about 39.4 ms), the part of the short-circuit current flowing through L_{limit} is not commutated. Normally, the current through L_{limit} is supposed to circulate through TSG_2 until it is damped out by the inherent resistance in the current path. However, instead the current through L_{limit} flows through the L_{stray} setting up a current path through $\text{TSG}_1 \rightarrow \text{AB}_1 \rightarrow L_{\text{limit}} \rightarrow L_{\text{stray}}$. In hindsight it is an important lesson that the arcing prolongation circuit must not be placed between the power source and the overcurrent protection circuit. Similar phenomena could lead to DC current conduction of the transformer secondary winding[9].

Depending on the test setup, the stray inductance can increase significantly during the actual testing. In fact as stray inductance exceeds a few tens of μH, an instant commutation becomes challenging even if the arc voltage of AB_1 exists. A possible way to enhance commutation is to use a double chamber interrupter AB_1 – see Figure A.8. The latter has increased arc voltage and was demonstrated to result in an improved commutation. Moreover, as the rating of the DUT increases some practical challenges associated with the arcing time prolongation circuit arise. This is discussed in detail in the appendix A.6.1

8.2.4. DC recovery voltage application

When in service, the HVDC CB is subjected to the system recovery voltage immediately after current suppression. In a test circuit supplied by an AC power source, the DC recovery voltage must be provided by a separate source. Using a similar approach as in a synthetic testing of HVAC CBs, a DC recovery voltage can be applied, for example, from a charged capacitor bank as shown in Figure 8.2. However, unlike in HVAC CB testing where the moment of current zero can be determined from circuit parameters and the power frequency, in HVDC CB testing the instant, at which the DC voltage should be applied, cannot be precisely determined in advance[10] – since the latter depends not only on the parameters of the test circuit but also on many parameters of the DUT. Even if the current zero can be determined, the application of DC recovery voltage presents many practical challenges. One of the main challenges is how to ensure a seamless transition from the TIV to the DC recovery voltage while maintaining the voltage stress and minimizing the current exchange. Also, a single approach may not be applicable for all the technologies of HVDC CBs – a customized solution depending on the type of the DUT might be needed.

Depending on the availability of components at a test facility, some technical solutions can be used. For example, a DC recovery voltage can be applied from a pre-charged

---

[9] L_{limit} could link with the transformer secondary winding in a similar way as it links with the stray inductance in the current position.

[10] only by accurate simulations

capacitor using a stack of high-voltage thyristors[11]. In order to maintain DC voltage application for sufficiently long duration, high-frequency triggering pulses are continuously applied to the thyristors. In the absence of a stack of thyristors, a stack of diodes in series with a mechanical closing switch can be used. The mechanical switch is pre-tripped in such a way that its contacts touch (pre-strike) during the last phase of the current suppression period – the precise timing is not critical as long as it falls within this period. The diodes are needed to prevent the charging of the DC recovery voltage source capacitor by the TIV of the DUT. In addition, the diode stack ensures a smooth transition from the TIV to the DC recovery voltage naturally when the former falls below the latter. However, this approach is challenging when the current suppression duration is too short, for example, during a test, in which low energy is absorbed. In such a case the scattering of the closing time of a mechanical switch may result in the start of conduction outside the current suppression period. The other drawback of the latter approach is that the DC recovery voltage is applied under any circumstance – the mechanical switch is tripped in advance regardless of whether a DUT interrupts or not. Thus, the diode stack must be capable of handling the short-circuit current due to the discharge of the DC recovery voltage source capacitor in case the DUT fails to clear or remains in a closed position for whatever reason.

In the absence of the above PE switches, this study proposes a method based on commonly available components at a high-power test laboratory. The proposed method uses a parallel combination of TSG_3 and a mechanical switch as shown in Figure 8.2. The approach is based on the fact that all HVDC CB technologies constitute capacitors either in the form of snubber circuits and/or as pre-charged capacitors for counter current injection – see details in Chapter 3. This capacitor is charged to the same voltage as the TIV during fault current suppression. Hence, the DC recovery voltage can be applied by creating interaction of this capacitor with the capacitor of a DC recovery voltage source.

The DC recovery voltage is applied first by triggering TSG_3 for precise timing[12]. This causes a charge balancing current to flow from the DC recovery voltage source to the DUT or vice versa depending on which capacitor is charged to a higher voltage at the end of the current suppression period. To ensure DC voltage stress when the charge balancing current ceases to flow through TSG_3, a pre-triggered HVAC CB AB_2 in parallel with TSG_3 is closed to bypass TSG_3 shortly after triggering. Moreover, the duration of charge balancing current can be prolonged by adjusting R_{DC} and L_{DC}.

Based on this concept, two solutions – with and without requiring real-time current

---

[11]Thyristors can be triggered within a few to tens of micro-seconds and the triggering signal is sent just before the end of the current suppression.

[12]Especially if the fault current suppression period is extremely short. If the current suppression time is long enough to accommodate the scattering of AB_2, then TSG_3 is not necessary.

zero detection – are demonstrated in the test laboratory. The schematic of the test circuit used for the demonstration of DC recovery voltage application is shown in Figure 8.7. The DUT is represented by a capacitor C_{DUT}, whereas the capacitor C_{DC} represents the DC recovery voltage source and is initially pre-charged. C_{DUT} is varied according to the equivalent capacitance of a DUT. This setup is sufficient for the purpose of demonstrating the DC recovery voltage application.

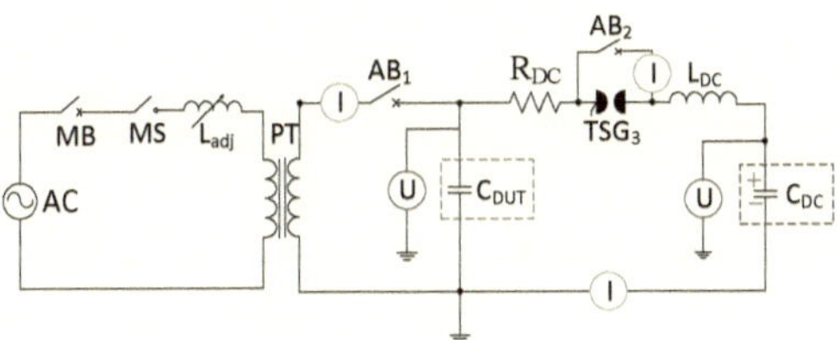

Figure 8.7.: Schematic of the test circuit for demonstration of the DC recovery voltage application method. ©[2019]IEEE.

Figure 8.8 shows test results demonstrating the method of applying DC recovery voltage after current suppression. In order to mimic the actual situation during current suppression, a small capacitive load current is interrupted by AB_1 – see Figure 8.7. When the current is interrupted by AB_1, C_{DUT} is charged to the peak value of the AC source voltage, which is 100 kV. This is assumed to correspond to a value of a TIV of an HVDC CB unit at the end of the fault current suppression time. In addition, the test was combined with a real-time current-zero detection where the interruption of the short-circuit current from the generator is detected.

Hence, upon the detection of the current suppression (at 5 ms in Figure 8.8), the TSG_3 is triggered. This results in current conduction of TSG_3 due to the voltage difference between the two capacitors. This current is shown by the red trace labeled as TSG_3 current in Figure 8.8. The DC voltage source capacitor (C_{DC}) is pre-charged to 90 kV[13]. A relatively large L_{DC}[14] is used to limit the oscillating current between the two capacitors, C_{DUT} and C_{DC}. The duration of current conduction through a spark gap very much depends on the current (as a result of the differential voltage), the gap length of the spark gap itself and the impedance between the two capacitors as well as the differences in

---

[13] A value chosen for demonstration purpose

[14] L_{DC} is chosen to be large enough to limit current flowing into the capacitor bank of the DC voltage source in case AB_1 fails to isolate the generator side from the rest of the circuit.

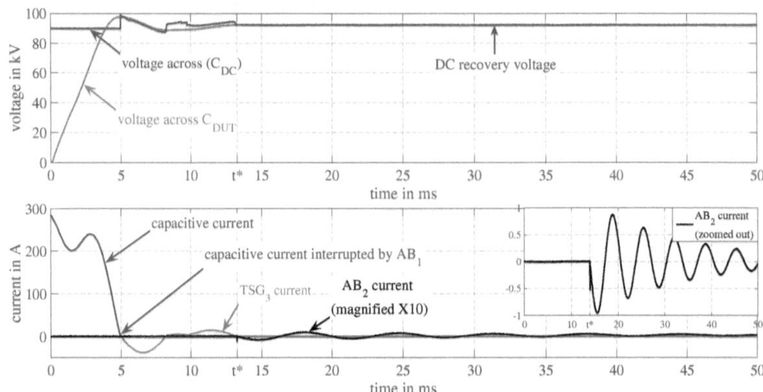

Figure 8.8.: Experimental demonstration of DC recovery voltage application right after current interruption. ©[2019]IEEE.

capacitance of the two capacitors[15]. During the arcing of TSG_3, the auxiliary breaker AB_2 is tripped in advance (see the timing diagram in Figure 8.1), and its contacts close at t^* in Figure 8.8. Then, a small residual charge exchange current (<1 A) continues to flow as shown in the figure (magnified tenfold for visibility). The actual current measurement is shown in the zoomed plot in the bottom graph of Figure 8.8.

Figure 8.9 shows the second method of applying the DC recovery voltage – without current zero detection. In this case the TSG_3 can be triggered during the current suppression period – preferably during the last stage of this period. The spark gap can be triggered from a sequence timer; hence, there is no need for real-time current zero detection. Also, there is no need for capacitive current interruption in this demonstration. Rather, the two capacitors, C_{DUT} (6 μF) and C_{DC} (10.7 μF) are pre-charged to 222 kV and 160 kV, respectively. R_{DC} (400 Ω) and L_{DC} (68 mH) are chosen so that the discharge from C_{DUT} to C_{DC} is over-damped. The discharge current[16] from C_{DUT} charges C_{DC} to a slightly higher voltage. However, the final DC recovery voltage is 180 kV, which is the desired level for the DUT under consideration. During the experimental investigation

---

[15]The resulting equivalent capacitance must not be too small in order not to lead to a high-frequency damped oscillation.

[16]This current has negligible effect on the total suppression current of the DUT.

it was observed that up to 10 ms current conduction duration of TSG_3 can be achieved using such a method. This method has been found convenient and was employed in the actual testing of the prototype HVDC CBs.

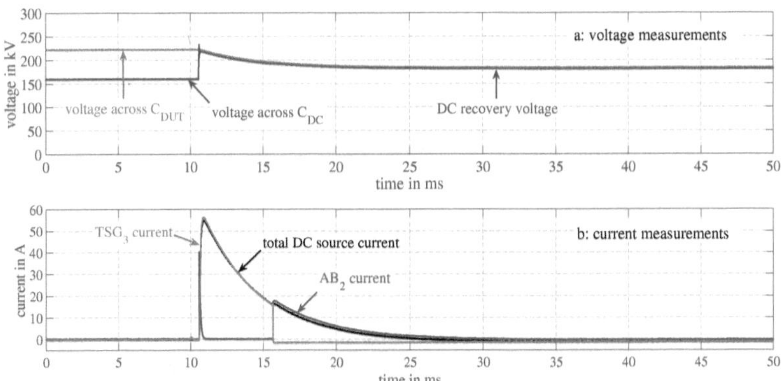

Figure 8.9.: Demonstration of DC recovery voltage application during current suppression

In general, given the source impedance, the discharge current may not be so significant if the DUT fails to sustain the DC recovery voltage for a sufficiently long duration. However, for a test result to be considered successful, the DC recovery voltage must not fall below a certain threshold for at least 300 ms, which is considered as a reasonable re-closing time or until a residual current breaker opens to isolate the HVDC CB from the system.

8.3. Performance Demonstration of the Test Circuit

In this section, the correct functioning of the complete test circuit, which was discussed in the previous section, is verified and its performance is evaluated. This is carried out by actually performing tests on prototypes of various technologies of HVDC CBs supplied by OEMs. While the performance of the HVDC CBs can be evaluated from these tests, the main focus in this section is the evaluation of the performance of the test circuit; whether the method can deliver the required stresses, whether all the parts of the test circuit can function as desired and operate in combination as a whole, and to identify practical limitations/challenges of the method and the circuit.

The performance of three different technologies of HVDC CBs – active current injection based on the direct discharge of a pre-charged capacitor, active current injection based on the VSC assisted resonant current (VARC) and the hybrid technology based on the original topology – have been demonstrated using the test method and the circuit developed in this book. The main components, electrical design and detailed operation principles of these HVDC CB concepts are discussed in Chapter 3.

8.3.1. Test procedure

An HVDC CB must be tested for the worst-case situations they could face in service. The worst-case situation may not necessarily be only the interruption of the maximum breaking current. Although it depends on the HVDC CB technology, the interruption of low current does not mean that it is easy. Therefore, HVDC CBs must be tested not only for the maximum current breaking capability but also for range of currents up to the maximum current breaking capability. For this reason, a test circuit also needs to supply a range of currents from low up to the maximum interruption capability. Initially, four test duties following the AC CB standard were defined: TF100, TF60, TF30 and TF10, corresponding to 100%, 60%, 30% and 10% of the maximum breaking current, respectively. The test results are discussed in detail in [BPS19, BPS+18a, BSN20].

Later, it was found out that the most critical test duties are the low current and the maximum current interruptions. After several test campaigns, it was clear that the intermediate test duties do not provide additional information. All the necessary stresses during the intermediate current interruptions are covered by the extreme duties. For this reason, the test duties are further refined to TF100, TC100 and TC10 which are the rated maximum fault current, the rated load/continuous current and 10% of the rated continuous current[17], respectively. Table 8.1 illustrates the refined and agreed upon test duties.

Table 8.1.: Test duties of HVDC CB

Test name	Breaking current	Remark[18]
TC10	10% rated continuous current	Low current interruption capability
TC100	100% rated continuous current	Load/continuous[19] current interruption capability
TF100	100% rated peak fault (maximum) current interruption	Maximum current interruption capability and rated energy absorption

---

[17]This was discussed and agreed with the manufacturers within the project consortium.

Moreover, in the case where the TF100 test cannot be performed at the rated energy absorption either due to the limitation of a test circuit or if the DUT supplied is designed for lower energy absorption than rated, the multi-part testing proposed in Section 7.4 is performed. The multi-part testing alternative splits TF100 into TF100* and TDT. The TF100* demonstrates the rated maximum current interruption capability at reduced energy whereas the TDT subjects the HVDC CB to the rated TIV duration at reduced energy. Table 8.2 illustrates the multi-part testing equivalent of TF100 test duty.

Table 8.2.: Equivalent multi-part test of TF100

Test name		Breaking current	Remark[20]
TF100	TF100*	100% rated peak fault (maximum) current interruption	maximum current interruption capability and reduced energy absorption
	TDT	Determined based on specified energy and rated duration of the TIV	Rated fault current suppression duration and reduced energy absorption

8.3.2. Test setup 1 (80 kV active current injection HVDC CB)

A photo of the test setup of the 80 kV prototype of the active current injection HVDC CB is shown in Figure 8.10. The components of the HVDC CB include a high-voltage vacuum circuit breaker (VCB) and a high-speed vacuum making switch (HSMS) contained in one enclosure, current injection capacitor bank and reactors, MOSA and a control cubicle. A current zero in the VI is created by closing the HSMS. In this case, the target was to achieve current zero in the VI within 8 ms after receiving a trip command [BPS$^+$18a].

The triggered spark gap TSG_1 and the auxiliary SF_6 AC CB AB_1 shown in Figure 8.10 are part of the test circuit, not the DUT. The remaining parts of the test circuit are not shown in the photo. In this test setup, the parts of the test circuit including arcing time prolongation and DC recovery voltage application circuits are not included. However, the overcurrent protection circuit has been implemented as shown in the figure. In other test setups, the complete test circuit is demonstrated and discussed along with the test results.

---

[18]Tests are performed per current direction in case the HVDC CB has bidirectional current interruption capability.

[19]In service condition, the terminal-to-earth voltage of the HVDC CB during the continuous current interruption can be as high as 2.5.p.u. since the breaker builds the TIV (1.5 p.u.) on top of the system voltage (1.0 p.u). However, in this test campaign the terminal-to-earth dielectric withstand is not considered since one side of the breaker is earthed.

[20]Tests are performed per current direction in case of bidirectional HVDC CB.

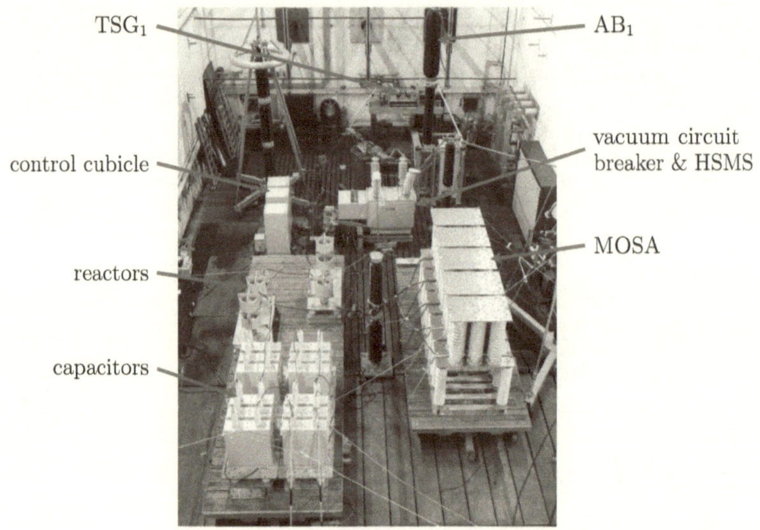

Figure 8.10.: Test setup of prototype active current injection HVDC CB tested at KEMA Labs [by kind permission of MITSUBISHI Electric Europe]©[2020]IEEE.

8.3.2.1. Test results

In this test setup the impact of the test circuit parameters, specifically the impact of the source voltage amplitude on the energy absorption, is demonstrated. For the 80 kV active current injection prototype, the test circuit parameters are designed to supply energy not exceeding 1.5 MJ to perform as many as possible tests in the available time. Otherwise, the multiple short-interval energy injection leads to accumulative temperature rise. Following the discussion in Section 7.3, the necessary test circuit parameters (the source voltage, circuit inductance and the making angle) are computed. Assuming 2 ms of relay time and 8 ms of the breaker operation time, a source voltage with a peak value of 19 kV is calculated to deliver 1.5 MJ. Later, to demonstrate higher energy absorption, for example 4 MJ, a source voltage amplitude is raised to 40 kV while increasing the circuit inductance proportionally to keep the test current to the same level. The test circuit parameters for the two different energy absorption at TF100* test are shown in

Table 8.3. In both cases a making angle of 39° relative to the source voltage zero is used.

Table 8.3.: Test circuit parameters for TF100* duty at different energy absorption

Test name	Inductance (mH)	Current (kA)	Energy (MJ)	Source voltage peak (kV)
TF100*	10.5	16	1.5	19
TF100*	20.5	16	4	40

Figure 8.11 shows test results of TF100* (16 kA) current breaking both in the forward and reverse direction. The prospective current is superimposed (shown by the dashed black traces in the top graphs) to provide a reference for comparison. In both tests, the current injection capacitor of the DUT is charged at the same polarity[21]. The current through the VI of the DUT is depicted in each graph to show local current interruption. As can be seen from Figure 8.11c, in the case of reverse current interruption, the injection current is superimposed onto the short-circuit current through the VI in the first half cycle and then creates current zero in the next one.

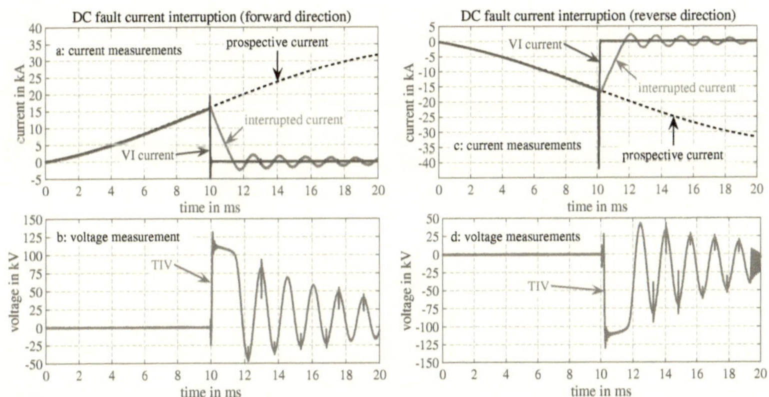

Figure 8.11.: 16 kA (TF100*) bidirectional DC short-circuit current interruption test results of an 80 kV active current injection HVDC CB prototype. ©[2019]IEEE.

During both the forward and reverse current interruption tests, the DUT suppresses the

---

[21]i.e., the superimposed high-frequency injection current starts with the same (negative) flow direction in both cases (can be seen in the current measurement)

current from just over 16 kA, which would otherwise increase to a peak value of 33.5 kA. The traces in the Figure 8.11b) and d) show the TIV generated by the DUT during the fault current suppression time. The high-frequency oscillation observed at the peak of the TIV is due to stray inductance in the loop between the capacitor and the MOSA of the DUT. In both cases the breaker absorbed nearly 1.4 MJ energy from the test circuit. Since at this stage the main focus is on the current interruption, the AB_1 is not tripped and hence, the DC recovery voltage after the current suppression is not applied. Therefore, after current suppression, there is an oscillation of a few hundreds of hertz due to the interaction between the charged capacitor of the DUT and the inductance of the test circuit.

Next, higher energy absorption is demonstrated. Figure 8.12 shows the oscillograms of another TF100* test, in which nearly 4 MJ energy is absorbed. This is achieved by adjusting the test circuit parameters as shown in the Table 8.3. In this case, AB_1 is

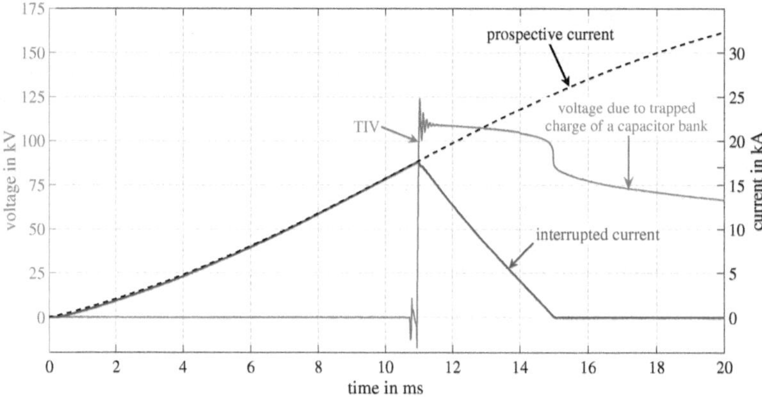

Figure 8.12.: Demonstration of 4 MJ energy absorption when interrupting TF100* forward current

tripped although there was no DC recovery voltage applied from the test circuit. However, a decaying DC voltage due to trapped charge on the DUT's injection capacitor is observed after the current suppression. Initially, this was considered to supply the DC recovery voltage; however, this is against the best test practice since this stress is produced by the DUT itself and not by an external circuit as in service. Nevertheless, it can be seen that AB_1 is providing galvanic isolation[22] between the power source and the DUT, thus making it ready for DC recovery voltage application as described in Section 8.2.

An important observation from the test result of Figure 8.12 is the duration of the TIV. The TIV of more than 110 kV is maintained for about 4 ms, thus stressing the DUT as well as the test circuit components for the same duration.

8.3.3. Test setup 2 (160 kV active current injection HVDC CB)

In the next test setup, the operation and performance of the complete test circuit is discussed. The photo of the test setup is depicted in the Figure 8.13. This is another prototype of the active current injection HVDC CB rated for 160 kV system voltage and 16 kA – TF100. On the left side of the figure, the components of the laboratory prototype installation of the active current injection HVDC CB are shown, whereas on the right side, the components of the test circuit are shown as labeled.

Figure 8.13.: Complete test setup of a 160 kV active current injection HVDC CB [by kind permission of MITSUBISHI Electric Europe].

In order to double the system voltage to 160 kV (based on the 80 kV prototype), two series connected VCBs are used in the CCB while two series connected vacuum HSMSs

---

[22]There is no oscillation observed after current suppression as in Figure 8.11.

are used in the current injection branch. Before each test, the counter current injection capacitor is pre-charged to -160 kV.

8.3.3.1. Test results

Figure 8.14 depicts the TF100* test result of the 160 kV prototype of the active current injection HVDC CB. In this test, all the features and sub-circuits of the complete test circuit are used. The arcing time prolongation circuit is utilized, by which additional 7 ms of current conduction duration is achieved in AB_1 before the actual short-circuit is applied. The overcurrent protection circuit is set to 20 kA, and the AB_1 is tripped to provide galvanic isolation of the power source and the DUT after the current suppression.

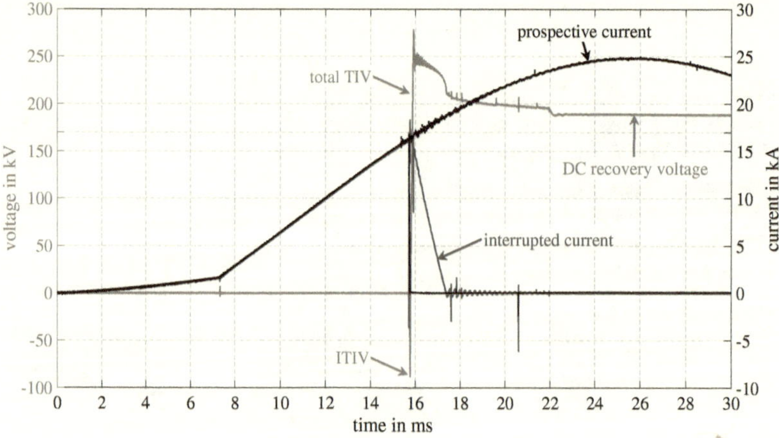

Figure 8.14.: Test result of 160 kV active current injection HVDC CB – 16 kA (TF100*) current interruption

The breaker suppresses a rising short-circuit current at 16 kA within the breaker operation time of 7 ms. The prospective current with a peak value of 25 kA is superimposed in the oscillograms to provide a reference for comparison. The breaker produces a TIV with a peak value of 277 kV, which oscillates around 245 kV, while suppressing the short-circuit current within 1.75 ms and absorbing nearly 3.1 MJ of energy. After the current suppression, the DC recovery voltage of 180 kV[23] is applied from the test circuit for over 1 s[24].

The current interruption becomes challenging as the test current is reduced to (and below) the rated continuous current. This is because the initial TIV (ITIV) due to the residual charge across the current injection capacitor at the moment of the local current zero in the VIs increases as the interruption current decreases. This can be observed from Figure 8.15 showing the test result of TC100 although re-ignition occurs before the full ITIV appears across the VIs at each current zero[25]. Nevertheless, the VIs sustained just over 110 kV ITIV at the last current zero in this case. Upon the local current interruption, this voltage appears immediately across the VI and becomes even higher during lower current interruptions, for example, during TC10 interruptions. This observation is, in fact, the rationale behind including the low current test duties (TC10 and TC100).

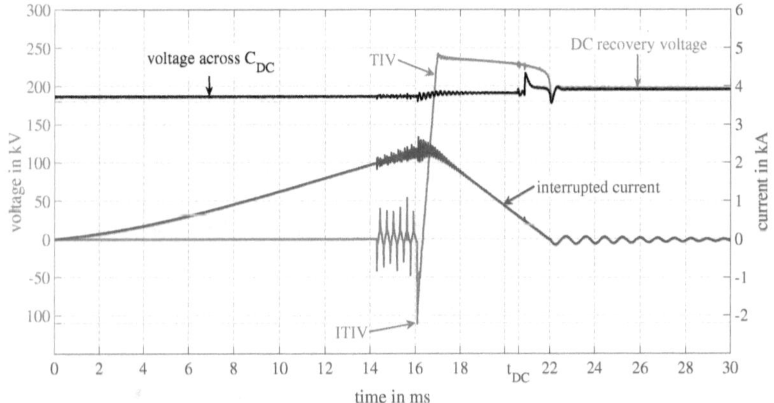

Figure 8.15.: Test Result of 160 kV active current injection – 2 kA (TC100) current interruption followed by DC recovery voltage application

---

[23] 180 kV DC recovery voltage is applied assuming maximum continuous system overvoltage of 10-15% above the system nominal voltage (160 kV in this case).

[24] It was agreed to apply DC recovery voltage for a duration of only 300 ms. However, to avoid risk of damage (long arcing) to the AC CB (used as disconnector) due to DC leakage current through MOSA, it was decided to keep the recovery voltage for longer duration until the discharge (earthing) switches are closed.

[25] The re-ignition phenomenon is discussed in detail in Chapter 9.

Another point of particular interest in this test is that the DC recovery voltage is applied 1 ms before the completion of the fault current suppression – as discussed in Subsection 8.2.4. This is marked as t_{DC} in Figure 8.15 and is achieved by closing AB_2 in advance so that its contacts touch at this moment. The voltage level of the DC recovery voltage capacitor C_{DC} is shown by the black trace in the figure. It can be seen that from t_{DC} onward C_{DC} is charged to a slightly higher voltage (200 kV) by the short-circuit source of the test circuit because the TIV of the DUT is higher then the pre-charge voltage (180 kV) of the capacitor[26]. As a result at the end of the current suppression period the DC recovery voltage of 200 kV is applied and maintained for about 1 s[27].

8.3.4. Test setup 3 (hybrid HVDC CB)

A prototype of a hybrid HVDC CB discussed in Section 3.4.1 is also tested using the complete test circuit. The detailed electrical diagram of the test circuit and the prototype installation test setup together with the test circuit components are shown in Figure A.7 and A.8 in the appendix, respectively. The prototype breaker is rated for 350 kV system voltage.

8.3.4.1. Test results

Figure 8.16 shows a test result of the prototype hybrid HVDC CB, in which the complete test circuit is used. The test result depicts that the prototype breaker suppresses 20 kA current in the test duty TF100* while producing a TIV of nearly 490 kV. The prototype breaker produced the TIV after the breaker operation time of 3 ms. It can also be seen that a 380 kV DC recovery voltage (assuming 10% rated continuous overvoltage) is applied after current suppression for about 1 s. The breaker suppresses the fault current from 20 kA to a leakage current level within 1.5 ms. The breaker absorbed energy of 6.5 MJ although the test circuit was designed to deliver 10 MJ energy. The latter is one of the challenges of this test method when testing HVDC CBs. The reason for the test circuit not delivering the targeted energy is discussed along with a few other test results in the next section.

Furthermore, a test result of the hybrid HVDC CB prototype, in which a 9.2 MJ energy was absorbed while breaking 16 kA is illustrated in the Figure A.9 in the appendix.

---

[26]Care must be taken here by placing proper impedance in the DC recovery voltage source circuit. Otherwise, the C_{DC} could be charged to very high voltage.

[27]In fact there is leakage current through MOSA resulting in 177 kV recovery voltage after 750 ms.

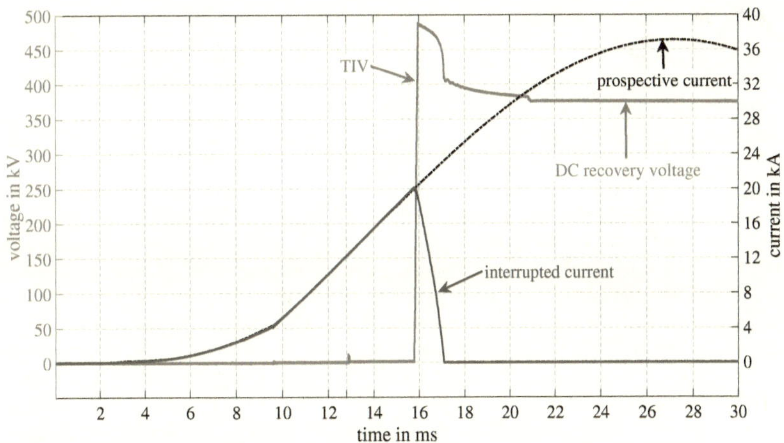

Figure 8.16.: Test result of a 350 kV hybrid HVDC CB – TF100* (20 kA) current interruption. 380 kV DC recovery voltage was applied.

8.4. Challenges of Testing Multi-Module/Full-Pole HVDC CB

The test results of the HVDC CB prototypes reported in the technical literature are mainly proof of industrial concepts based on single modules rated for $80 - 120$ kV. However, it is essential to verify the performance of full-pole (full-scale) equipment. A few tests of full-scale product have been reported recently [ZWZ+15, TWZ+, TZH+, ZXM+18, GFG19, ZYZ+20]. However, due to apparent limitations of the test circuits, tests have been performed at much lower energy than the rated value.

There are two main requirements which a test environment must fulfill for the complete performance demonstration of a multi-module/full-pole HVDC CB. First, it needs to have sufficient short-circuit power to supply the rated stresses: current, energy and voltage. Second, and a unique challenge of testing HVDC CB is that the test installation should be able to withstand the TIV generated by a full-pole HVDC CB. These challenges are discussed below.

8.4.1. Short-circuit power

This requirement can be met by availability of, for example, multiple short-circuit generators and step-up transformers. Nonetheless, a lot depends on the system as well as HVDC CB parameters, for example, the fault neutralization time, the maximum current interruption capability and the energy rating. At a given power frequency, the fault neutralization time along with the maximum current interruption capability dictate the di/dt of the source current, whereas the energy rating dictates the magnitude of the source voltage that a test circuit must supply, as discussed in Section 7.3.

Considering the actual installation at KEMA Labs with up to six short-circuit generators (of 2250 MVA each at 50 Hz) and ten step-up transformers (550 kV AC insulation level), the maximum energy that can be supplied to different technologies of HVDC CB at different rated voltages is depicted in Figure 8.17. Four HVDC CB technologies with fault neutralization times of 3, 5, 7 and 9 ms, and capable of interrupting 16 kA are considered.

It can be derived from Figure 8.17 that a source current with a di/dt in the range of $1.8 - 5.3$ kA/ms can be achieved[28]. Except in some cases (at higher rated voltage) where it is not possible to achieve the required di/dt with a desired source voltage[29], the peak values of the source voltage are adjusted to the same value as the CB rated voltage. The figure shows that the maximum energy that can be supplied to the HVDC CB depends on both the fault neutralization period and the CB rating. For example, at rated CB voltage of 320 kV, up to 30 MJ can be supplied to an HVDC CB with fault neutralization time of 9 ms, whereas up to 15 MJ can be supplied to a CB with fault neutralization time of 3 ms. In general, the longer the fault neutralization period, the higher the energy that can be supplied[30]. If the short-circuit power source cannot supply the required current at the rated CB voltage (and rated energy), multi-part testing can be performed, in which TF100 is split into TF100* and TDT as shown in Table 8.2.

8.4.2. TIV withstand

The second main challenge is the impact of the HVDC CB TIV on the test installation – on the step-up transformers and the primary side installations. If the peak value of the source voltage is much lower than the rated voltage of the CB, it results in a smaller necessary transformer ratio and hence, a fewer series connected transformers than would be required

---

[28] Prospective currents at higher di/dt up to 10 kA/ms have been demonstrated at shorter fault neutralization periods, but not relevant considering the recent performance of HVDC CBs.

[29] due to quadratically increasing circuit impedance as result of increased transformer ratio

[30] This is because, at a given voltage, circuit inductance must be increased to decrease the rate-of-rise of current over a prolonged fault neutralization period.

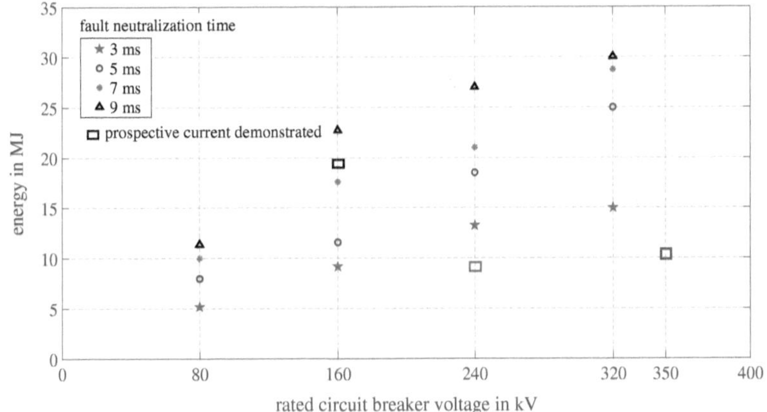

Figure 8.17.: Maximum energy supply versus voltage ratings considering different technologies of HVDC CBs at KEMA labs. For all technologies rated short-circuit breaking current of 16 kA is assumed. The rectangles show prospective current demonstration considering a technology identified by the color. ©[2019]IEEE.

for rated voltage. The TIV produced by the HVDC CB is, however, independent of the source voltage, which drives the short-circuit current. The TIV appears also across the transformer secondary winding and is transformed (according to the transformer ratio) to the primary side of the test installation.

There are two consequences of the transformation of the TIV to the primary side. First, because the transformer ratio is primarily chosen to obtain the desired source voltage without considering the TIV, the stepped-down TIV appearing on the primary side could result in an overvoltage to the components and busbars. This is mostly an issue when testing EHVDC CBs at significantly lower energy than the rated value. Thus, the transformer ratio must not only be selected to get the desired source voltage for a specific test but also the choice must consider the resulting TIV on the transformer primary side. To avoid this risk, a minimum transformer ratio must be ensured when designing a test circuit for a specific test as in Equation (8.1).

$$N_{\mathrm{T}} \geq \frac{U_{\mathrm{CB,pk}}}{U_{\mathrm{LV,pk}}} \tag{8.1}$$

Where $U_{\mathrm{CB,pk}}$ is the peak value of the TIV, $U_{\mathrm{LV,pk}}$ is the peak withstand voltage of the primary side installation and N_{T} is the transformer ratio.

The minimum transformer ratio entails a minimum circuit impedance. Moreover, in order to obtain the test current, the selection of the source voltage magnitude must consider the circuit impedance – the combination of which would lead to a minimum energy absorption. In conclusion, based on Equation (8.1), the possible energy absorption of an HVDC CB cannot be arbitrarily low. Therefore, for an HVDC CB with a known TIV, there is a minimum energy absorption required to ensure safe voltages on the primary side equipment.

Second, the TIV might push the transformer into saturation. This is discussed in the following section.

8.4.3. Transformer saturation

The other issue, which is the main challenge of the AC short-circuit generator based test method, is the transformer saturation. Normally a transformer saturates when excessive voltage and/or a prolonged duration unipolar voltage is applied to the primary winding. When testing an HVDC CB, the TIV appears on the secondary windings and thereby induces additional magnetic flux[31] in the core. As a result, it appears as if the transformer is energized from the secondary side during the current suppression period. Since, at the transformer terminal, the TIV is of the same polarity as the source voltage, magnetic flux proportional to the difference between the TIV and the source voltage is additionally induced. Especially during a test with long duration TIV, the additional flux can be significant[32]. Later, when the source voltage changes polarity, the additional magnetic flux induced due to the TIV is not compensated. This is illustrated in Figure 8.18. The figure shows the voltage measurement at the secondary side of the transformer during one of the tests of the 350 kV hybrid HVDC CB. The voltage measurement is located on the source side of the $\mathrm{AB_1}$ in Figure 8.2. Because the DUT acts as a source during current suppression, the impact of saturation is not yet observable until the next cycle.

Moreover, for the purpose of arcing time prolongation, a short-circuit is applied at nearly 100% asymmetry. This, depending on the magnitude of the source voltage, accumulates significant magnetic flux in the core of the transformer already by the time the TIV is

---

[31]in addition to the flux induced by the source

[32]The actual magnitude depends on the voltage difference between the source and the TIV and the current suppression duration.

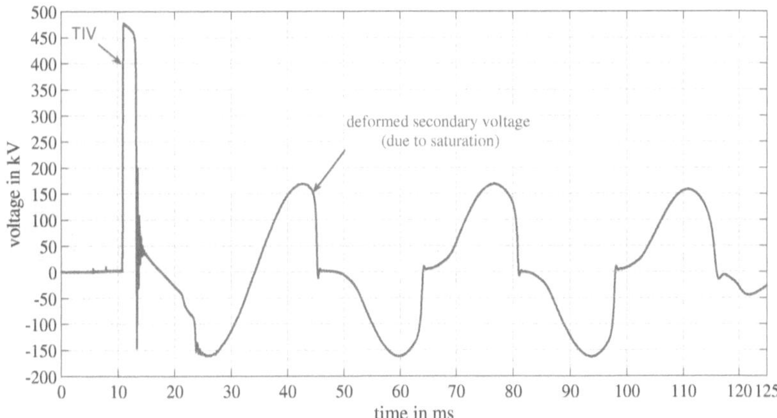

Figure 8.18.: Measurement of transformer secondary side voltage after saturation during current interruption by an HVDC CB.

generated. In addition, after current suppression, the transformer remains energized only for about $2-3$ cycles[33], which is not long enough for the asymmetric flux to disappear – see Figure 8.18. As a result, considerable magnetic flux remains in the core at the end of a test. Also, when testing unidirectional current interruption, tests are performed using unipolar source voltage successively. This leads to flux accumulation over successive tests unless some de-saturation steps are taken. This is described by Equation (8.2) below,

$$\Phi(t)_n = \int_{t_{MS}}^{\frac{T}{4}} \frac{\widehat{U}\sin(\omega t)}{N_1} dt + \int_0^{T_{FS}} \left[\frac{U_{CB}}{N_2} - \frac{\widehat{U}\sin(\omega t + \theta_2)}{N_1} \right] dt + \Phi_{r_{n-1}} \qquad (8.2)$$

Where $\Phi(t)_n$ is the magnetic flux induced during the nth test, T is the period of the power frequency, $t_{MS} = \frac{\theta_{MS}}{2\pi f}$ is the time measured from voltage zero till short-circuit making, $\widehat{U}\sin(\omega t)$ is the voltage applied at the primary winding, N_1 and N_2 are, respectively, the number of winding turns on the primary and secondary sides, T_{FS} is the fault suppression

---

[33]first through the MB, which opens at three quarters of a period, and later through a high impedance in parallel to the MB

duration, U_{CB} is the TIV produced by the breaker, θ_2 is the phase angle at the beginning of current suppression (as in Equation (7.3)) and $\Phi_{r_{n-1}}$ is the remanent magnetic flux in the core from the previous test or initially (if) available at the first test.

The first term on the right side of Equation (8.2) corresponds to the magnetic flux induced due to asymmetric making, whereas the second term on the same side corresponds to the additional flux induced by the TIV of the DUT. For a given magnetic core cross-section, a transformer saturates when $\Phi(t) > \Phi_{\mathrm{saturation}}$, i.e. when magnetic flux density B exceeds a saturation threshold ($B_{\mathrm{saturation}}$).

Several test results revealing the impact of the transformer saturation are obtained during the test campaign. Figure 8.19 shows test results of two such successive tests performed under identical circuit conditions. In these tests four short-circuit generators and six short-circuit step-up transformers[34] are connected in parallel and in series, respectively. In fact, multiple tests under the same circuit conditions, in which transformer saturation is observed, have been performed prior to these tests. These tests are chosen because the impact of transformer saturation on the TIV duration as well as the differences in the instant of the onset of saturation can clearly be seen.

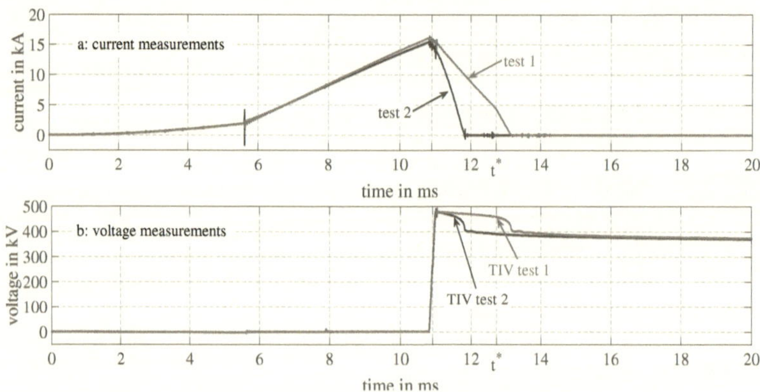

Figure 8.19.: Impact of transformer saturation during current interruption by an HVDC
CB – comparison of interrupted current and TIV measurements at two
successive tests

---

[34]with transformer ratio of 14/216

In each test case, the transformer saturation results in a different current suppression duration. This is because of the difference in the initial remanent magnetic flux ($\Phi_{r_{n-1}}$ in Equation (8.2)), which in turn results in different instants at which the transformers saturate. In the first test (shown by the red traces), the transformers saturate but at a later stage of the current suppression period – at t^* shown in the figure where the change in the rate of current suppression is seen in the current measurement. In this case the current suppression time is slightly reduced (by <200 μs) from the expected duration. As a result, reduced amount of energy is absorbed – 9 MJ instead of the expected 10 MJ, whereas in the following test (shown by the blue traces), the transformers saturate at an earlier stage of the current suppression period – on the rising edge of the TIV or just after the full TIV is established. In the latter test the current suppression time is 1 ms and the energy absorption is 3.7 MJ, which are significant deviations from the expected values – 2.5 ms and 10 MJ, respectively.

From the moment a transformer saturates, the magnetic energy stored in the primary side circuit inductance along with the electrical energy injected by the source can no longer be delivered to the DUT since at this stage the primary and the secondary windings are decoupled. In such a case, only the energy stored in the secondary winding, which now appears as an air core reactor, is absorbed by the DUT. This is illustrated in Figure A.20 where the impact of transformer saturation can also be observed in the generator (primary side) current and voltage measurements.

Some measures can be taken to reduce the risk of transformer saturation. For instance, before each test, transformers can initially be demagnetized by applying a gradually decreasing primary voltage for about 10 minutes. Moreover, for the purpose of testing a unidirectional HVDC CB in particular, saturation can be slightly pre-compensated. This is possible by initially energizing the transformer in opposite polarity (to that of the test voltage) so that the remanent magnetic flux will also have opposite polarity to the flux that would be induced during actual testing. This can be achieved by repeatedly applying unipolar voltage so that the remanent flux accumulates over successive shots. In addition, choosing a more favorable transformer ratio can reduce the induced flux per voltage.

The best solution to address the transformer saturation is reducing the TIV per transformer. This can be achieved by increasing the number of series connected transformers. For example, in a TDT[35] test of a hybrid HVDC CB, ten short-circuit transformers are used in series. In this case a source voltage peak (at secondary side) close to the rated voltage of the DUT (85%) is used, and a TIV duration of 11 ms is targeted. The test result is shown in Figure 8.16.

In this TDT test 4.6 kA current is interrupted by the DUT. A significantly higher source

---

[35]A complementary test to TF100* Table 8.1 that is intended to maximize the duration of the TIV.

voltage peak (290 kV with a transformer ratio of 14/612) is used together with a much larger circuit inductance (558 mH) compared to the 139 kV source voltage peak used along with a 57 mH circuit inductance in the tests shown in Figures 8.19 and A.20. The test results of Figure 8.20 show that a TIV of just over 460 kV is maintained for about 9 ms – 2 ms shorter than the targeted duration. In the graph, the primary current as referred to the secondary (see the black dashed trace) is superimposed for illustration. Nevertheless, the transformers saturate about 7 ms into the fault current suppression period – at 20 ms on the graph. From this moment on, the primary and secondary side currents are no longer related by the transformer ratio, and hence the current measurements on the two sides deviate as explained in Appendix A.6.4. As a result the secondary side current is suppressed at $t = 22$ ms while over 7 kA current per generator is still flowing on the primary side and subsequently a saturated transformer current flows, as can be seen from the figure in the black dashed trace.

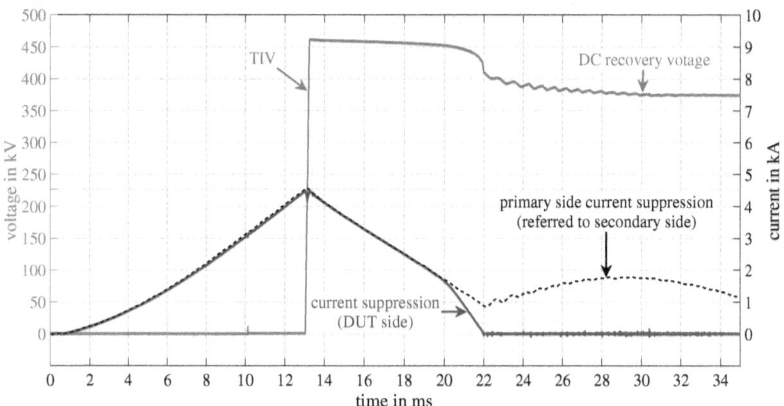

Figure 8.20.: TDT Test result of hybrid HVDC CB - 350 kV, 4.6 kA with DC recovery voltage application

In this case the transformers' saturation is caused by the long duration TIV – the duration in which the magnetic flux is almost linearly changing due to the plateau nature of the TIV. This is in fact a practical limitation of the test circuit to produce long duration TIVs.

The other best counter action to tackle transformer saturation is to perform tests at alternating polarity. This is specially convenient when testing bidirectional HVDC CBs. A reverse current interruption test can be performed by shifting the timing parameters of the test setup by 180°. This has been successfully demonstrated in the test laboratory. Even if the transformers saturate during current interruption in one direction, correct parameters can be obtained in the next test, in which the reverse current is interrupted, and no transformer saturation is observed in the latter. This can be understood from Equation (8.2) where the remanent flux from the previous test ($\Phi_{r_{n-1}}$) has opposite polarity to the flux produced during current interruption in the reverse direction.

8.5. Summary and Conclusion

The chapter presents the practical implementation of the complete test circuit that can verify the fault current interruption performance of HVDC CBs. The test circuit is based on AC short-circuit generators operated at low power frequency and synthetic DC recovery voltage application, both available in high-power laboratories. A pragmatic approach of testing HVDC CBs has been developed, realized and verified at a high-power laboratory using prototype HVDC CBs for testing and demonstration. Five industrial prototypes in the last phase of product development of three different HVDC CB technologies with ratings in the range from 25 kV to 350 kV have been demonstrated. In addition to the main short-circuit power source, many additional features and sub-circuits are needed to ensure the correct and safe operation of the test circuit while applying complete stresses in a single test. The additional sub-circuits for the protection of the DUT as well as the test circuit components and the correct and precise application of the DC recovery voltage have been implemented.

Based on the actual test results, the limitations of the test circuit have also been identified and methods to address some of these limitations are proposed. One of the limitations is to supply rated stresses to HVDC CBs with rated voltage above 500 kV and withstand the TIV even when, due to practical limitations, a test breaker is not equipped with the fully rated energy absorption module that is producing such overvoltage. A two part test procedure has been proposed to address the necessary stress except for the rated energy, has been proposed and demonstrated. The other critical challenge is the transformer saturation, as a result of which the desired test parameters cannot be achieved. Some practical transformer saturation mitigation techniques during HVDC CB tests have been demonstrated.

9. Experimental Investigation of Stresses on the Main Components of HVDC Circuit Breakers

This chapter discusses the important stresses on the major components of HVDC CBs for the purpose of justifying the test requirements defined in the previous chapters. Most of the results discussed in the chapter are published in [BSN20, BSN$^+$19]. The chapter is organized as follows. In Section 9.1, brief background of the study is presented. In Section 9.2, the test setup of the experimental DC CB along with its configuration and technical specification is presented. The actual test results of the three VIs are analyzed in detail in Section 9.3. The stresses of the MOSA are discussed in Section 9.4 together with a brief description of the MOSA module design for the HVDC CB. Finally the conclusion based on the results of the investigation is presented in Section 9.5.

9.1. Introduction

Unlike an HVAC CB, an HVDC CB is no longer a single device, rather it is a system of components designed and arranged in several current branches operating in a prescribed sequence to achieve DC current interruption as discussed in Chapters 2 and 3. These components include PE switches like IGBTs, BIGTs, IGCTs, IEGTs, thyristors, components such as MOSA and mechanical switching devices like UFD, VCB and HSMS including their newly developed high-speed actuators. Most of these are standard components used in non-standard applications, and therefore, have to face non-standard stresses [SB]. Some are newly developed, for example, the UFDs and the high-speed actuators on all other mechanical switching devices. In order to mitigate the risk of failure of these components under an unconventional application, thorough testing and investigation of stresses on these components under the new operation conditions is essential. In addition, to define and refine justified test requirements[1], a thorough understanding of the interactions

---

[1]The results are used as inputs to the ongoing standardization activities, which is an important spin-off of this study.

between the internal components of HVDC CBs under real DC fault current interruption
conditions is necessary. For this purpose, an experimental DC CB based on the active
current injection scheme is set up in a high-power laboratory. The performance of and the
stresses on the two main components of such a breaker, namely, the VI and the MOSA
are investigated.

The experimental DC CB uses VIs of commercial 36/38 kV AC VCBs as the main
interrupter. Three different designs of VIs are used in the investigation and it is found
out that each of the VIs behave completely differently. Moreover, during the experimental
test campaign, the stresses on MOSA modules, designed for HVDC CB application, are
also investigated by applying different amounts of energy per volume (70-230 J/cm^3) at
temperatures as high as 250 °C. In order to find out the performance limit of the MOSA
for this application, successive high-energy tests are performed until failure (electrical,
mechanical) occurs in the MO varistors. The detailed analysis of these failure mechanisms
during destructive tests and the root causes of these are presented.

9.2. Design and Specification of the Experimental DC CB

9.2.1. Experimental setup

The experimental DC CB is set up with a focus on the investigation of stresses on the
internal sub-components. The overall speed of operation of the experimental DC CB is
not a parameter of interest here. This is due to the fact that the mechanism/drives of
the VCBs used are not designed for DC CB application, rather for AC CB application, in
which much longer opening times (tens of milliseconds) are common[2]. The test parameters,
however, are similar to those used for testing a real (a scale model of) HVDC CB.

The electrical diagram of the test setup is shown in Figure 9.1. The main components
of the experimental DC CB can be seen in the dashed red box of the figure together
with the electrical measurements. It consists of a VI or series connected VIs in the CCB,
a pre-charged capacitor (C_{inj}), which supplies a counter injection current current zero
creation, an inductor in the current injection branch (L_{inj}), which is used to limit the
frequency and the peak value of the injection current, TSG$_2$ that serves as a high-speed
making switch and finally the MOSA, which is a crucial component for limiting and
maintaining the TIV across the DC CB during current suppression and hence, absorbing
the energy in the circuit.

Figure 9.2 shows a the photo of the laboratory test setup of the experimental DC CB

---

[2]Fast actuators based on Thomson coils (electromagnetic repulsion) are used in the latest HVDC CBs –
discussed in Chapter 3.

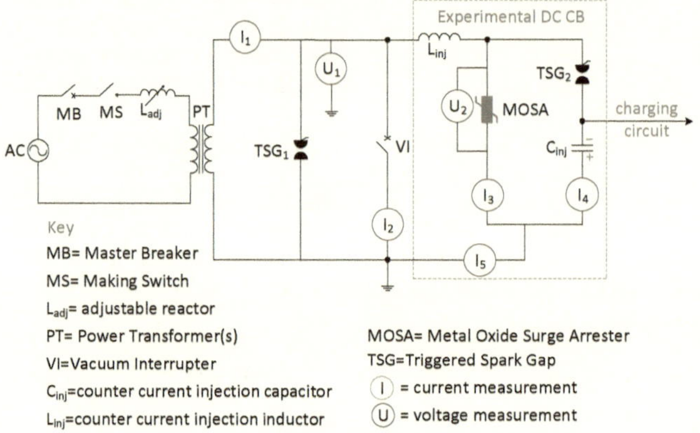

Figure 9.1.: Electrical diagram of the experimental DC CB test setup. ©[2019]IEEE.

with the main components labeled.

9.2.2. Technical specification of the experimental DC CB

The performance of a VI at different stages of the current interruption process depends on several factors such as the contact design, shape and size, contact materials and composition, gap distance, etc., [Sla07]. In order to investigate the impact of the VI contact design on the DC fault current interruption performance, the VIs of three different, standard off-the-shelf VCBs produced by various manufacturers are used. These are all three-phase AC CBs with ratings as shown in Table 9.1.

Table 9.1.: Specifications of VCBs used in the experimental DC CB

VCB Type	Rated voltage (kV)	Rated current (A)	Rated short-circuit breaking current (kA)	Opening time[3] (ms)	Magnetic arc control
A	38	2500	31.5	37.5	TMF[4]
B	36	2500	40	46.6	AMF[5]
C	36	2000	31.5	37	AMF

Figure 9.2.: Laboratory test setup of the experimental DC CB

In order not to exceed the AC voltage ratings of the VCBs, which is in the range of 36-38 kV (RMS) as shown in Table 9.1, the TIV of a single interrupter DC CB is set to 40 kV with a transient peak as high as 45 kV. In this case, only one out of the three poles VI is used while the remaining poles are not connected to the circuit. An electrical diagram of a single interrupter DC CB is shown in Figure 9.3a. In order to double the voltage rating of the experimental DC CB, two VIs (the peripheric poles) are connected in series, and to ensure the doubling of the TIV, two series connected MOSA modules are used as shown in Figure 9.3b. A DC fault current interruption of 16 kA is targeted in both cases. Nevertheless, the charging voltage as well as the values of C_{inj} and L_{inj} are adjusted in the double interrupter case such that the electrical stresses per component remain the same as for the single interrupter case. Moreover, in the double interrupter setup, there are voltage grading capacitors across each VI to ensure equal TIV distribution, see Figure 9.2 and 9.3b.

In the diagrams depicted in Figure 9.3a and 9.3b, the MOSA is connected directly in

---

[3] The time from trip command until contact separation

[4] Transverse magnetic field (also known as radial magnetic field)

[5] Axial magnetic field

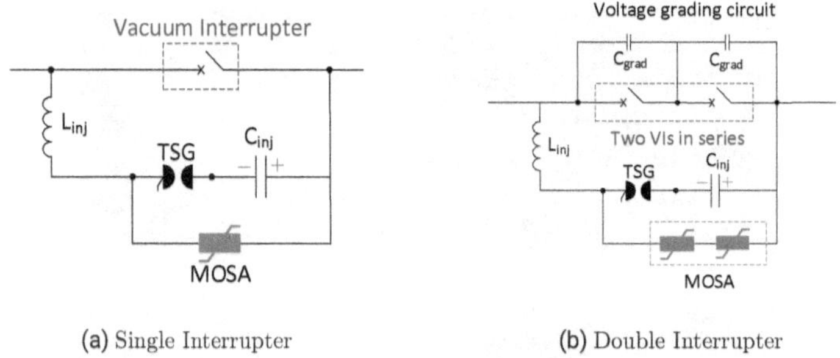

(a) Single Interrupter   (b) Double Interrupter

Figure 9.3.: Electrical diagram of experimental DC CB. ©[2020]IEEE.

parallel with C_{inj} and not with the VI(s). This is to reduce the voltage stress on the VI due to the transient overshoot during the current commutation from L_{inj} to the MOSA [BSN20]. In the layout shown in Figure 9.3, there is no need for current commutation from the L_{inj} to the MOSA during the entire current suppression period.

9.2.3. Design of the counter current injection circuit

The size and expense of the auxiliary components such as the injection capacitor are dictated by the performance of the VI, the system voltage and the current rating. Ideally, the commutation circuit is constructed with L_{inj}, and this in turn minimizes C_{inj} needed to produce the injection current [Pre82]. However, lowering L_{inj} and C_{inj} of the injection circuit raises the oscillation frequency. Although the high-frequency (HF) current interruption capabilities (di/dt and du/dt) of VIs are higher than for any other type of mechanical interrupters, there are limits [Sla07]. As these limits are approached, interruption becomes uncertain. Thus, the main question for the designer of a DC CB would be what is the minimum capacitance that results in a reliable interruption. This, in fact, is not the purpose of this investigation. Further details about this can be found in [Hei].

The design parameters for the current injection circuit components (C_{inj}, L_{inj}) are:

- Amplitude and frequency of the injection current

- Charging voltage of the capacitor (U_{ch})

The amplitude of the injection current is designed based on the maximum current interruption capability of the DC CB while considering a sufficient number of CZCs. It should be large enough to create a sufficient number of CZCs at the highest fault current. Its value is determined by the pre-charging voltage of C_{inj} and characteristic impedance (Z_c) of the injection circuit. The pre-charging voltage is normally equal to the rated voltage of the system, for which the breaker is designed. This is because it simplifies the charging of C_{inj} from the system DC bus. The injection current frequency should be high enough to create sufficient CZCs in quick succession[6] yet low enough to ensure that the VI can handle the di/dt at the CZCs. Using the characteristic impedance and the desired injection circuit frequency, the values of C_{inj} and L_{inj} are determined.

The current from the injection circuit is, therefore, given as:

$$i(t) = \frac{I_m}{k} \exp(-\gamma t/2) \sin(k\omega t) \tag{9.1}$$

where $I_m = \frac{U_{ch}}{Z_c}$ is the undamped amplitude of the injection current (assuming a lossless circuit), $Z_c = \sqrt{\frac{L_{inj}}{C_{inj}}}$, $k = \sqrt{1 - (\frac{R}{2Z_c})^2} \leq 1$ is the amplitude correction factor due to parasitic resistance R in the circuit, $\omega = \frac{1}{\sqrt{L_{inj}C_{inj}}}$ is the angular frequency of the injection current, $\gamma = \frac{R}{L_{inj}}$ and L_{inj} includes the stray inductance in the current loop.

When interrupting current I_{DC}, CZCs are created at every time when $i(t) = I_{DC}$ in Equation (9.1). The rate-of-change of current (di/dt) at any point on the injection current waveform can be obtained from the derivative of Equation (9.1) as follows,

$$\frac{di}{dt} = \left[\frac{U_{ch}}{L} \cos(k\omega t) - \frac{I_m \gamma}{2k} \sin(k\omega t)\right] \exp^{-\gamma t/2} \tag{9.2}$$

And, the voltage across C_{inj} at any time t ($U_C(t)$) after current injection is given by

$$U_C(t) = \left[U_{ch} \cos(k\omega t) + \frac{I_m R}{2k} \sin(k\omega t)\right] \exp^{-\gamma t/2} \tag{9.3}$$

The 2nd term in Equation (9.3) accounts for the charge loss (energy dissipation due to the inherent resistance) during the injection current oscillation.

From Equation (9.2), the di/dt is at its maximum near $t = \frac{\pi n}{k\omega}$ and at its minimum near $t = \frac{\pi(2n+1)}{2k\omega}$ where n is a positive integer. This means the di/dt at a CZC is higher when interrupting low current than when interrupting high current. Besides, the value of $U_C(t)$ in Equation (9.3) at each CZC represents the initial TIV (ITIV) across the VI due to remaining charge on C_{inj}. Similarly, the ITIV is at its maximum near $t = \frac{\pi n}{k\omega}$ and at its minimum near $t = \frac{\pi(2n+1)}{2k\omega}$. Thus, not only the di/dt but also the ITIV is large when

---

[6]without significantly increasing the internal current commutation time

interrupting low current.

For the experimental DC CB in this investigation, the pre-charge voltage of a single interrupter setup is $25 - 27$ kV. The peak value of the injection current is set to 20 kA at a frequency of $4 - 5$ kHz. Based on this information, the values of the injection circuit components and the expected di/dt while interrupting different current magnitudes are given in Table 9.2. The injection circuit parameters are designed so that at least 4 CZCs[7] can be created during the high-current test as illustrated in the next subsection.

Table 9.2.: Parameters of injection circuit and di/dt at different current interruption

Frequency (kHz)	$C_{\text{inj}}(\mu F)$	L_{inj} (μH)	di/dt $(A/\mu s)$ at the 1st CZC		
			2 kA	10 kA	16 kA
4	31.8	49.7	500	435	302
5	25.5	39.8	625	544	377

9.2.4. Experiment method and procedure

The electrical as well as thermal stresses seen by the VI depend on the magnitude of the fault current. With respect to the electrical stresses, it may not necessarily mean that a low current interruption is less severe than high current interruption for the reasons discussed in the previous subsection, whereas the arc thermal stress limits the maximum current interruption capability of a VI. In order to investigate the impact of the fault current magnitude on the performance of a VI, three test currents are defined as follows.

1. Low current: 2 kA

2. Medium current: 10 kA

3. High current: 16 kA

The other crucial parameter, of which the impact was investigated, is the arcing duration. The arcing duration is varied between $1 - 4$ ms. The arcing duration determines the contact gap-length although the latter is unknown due to a lack of information of the used VCBs. Moreover, because the details on other key design features, such as contact material composition, dielectric internal interrupter design etc., are not available, the conclusions drawn based on the observed experimental results are mainly qualitative.

---

[7]Due to inherent losses in the circuit, the injection current decays quickly, which limits the number of CZCs that can be created.

Experiments are repeated 10 times by keeping the parameters of interest the same, for example, the arcing duration of the VI(s).

Low frequency (LF) current is supplied by AC short-circuit generators as discussed in Chapter 8 and [BSTI16, BPS19]. The arcing prolongation and DC recovery voltage application sub-circuits are not included in this experimental campaign. An example of the prospective LF current produced during a test is shown in Figure 9.4 along with different timing signals. The experimental DC CB is operated in such a way that the contacts of the VI(s) are separated after the short-circuit current starts to flow. This means the VCB needs to be tripped at t_1, prior to the onset of the prospective current[8] which is at t_2. The breaker opening time, which is the duration from t_1 until t_3, is precisely known for the VCBs as shown in Table 9.1. Therefore, the trip command can be precisely sequenced in reference to the moment of short-circuit application as illustrated in Figure 9.4. From t_3 onward, the LF current flows through a vacuum arc shown by the thickened light red region. At t_4 the high frequency (HF) counter current is injected at a frequency of 4 kHz and a peak value of 20 kA ($\pm 5\%$).

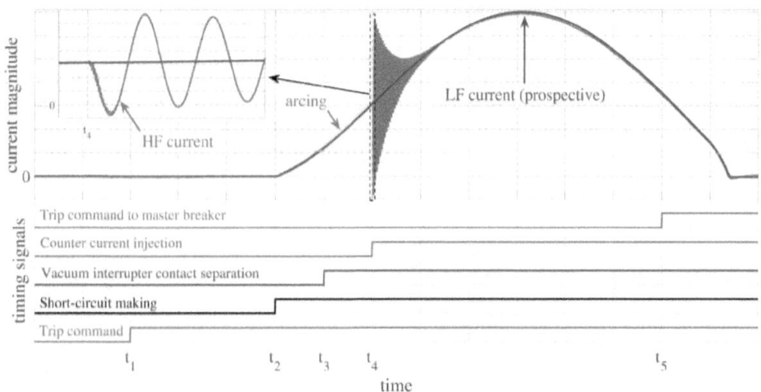

Figure 9.4.: Prospective current with counter current injection superimposed and timing diagrams ©[2020]IEEE.

In some literature a test is considered as failed if the VI does not interrupt at the first

---

[8]During the testing of the actual HVDC CBs, a trip command is sent after a short-circuit is applied to mimic the actual HVDC CB application.

current zero even if it cleared later after a few CZCs [Pre82]. The main motivation for this is that if a VI does not clear at the first current zero but at later current zeros, it shows that the conditions were near the limits of the interrupter. It is also to avoid EMI by the HF voltage spikes during re-ignitions. Then, it is much better to target the first CZC and consider the later CZCs as back up. From a DC CB designer point of view this is the preferred target; however, in this investigation a test is considered as a failed if the VI does not interrupt at any of the CZCs created in the process.

9.3. Experimental Results: HF Current Interruption Capability and Dielectric Coordination of VIs

9.3.1. Experimental results of VI type A

Figure 9.5 depicts typical 10 kA current interruption by a single VI of type A. Current and voltage measurements near CZCs are shown in the zoomed plots. The contacts of the VI separate at t_3. From this moment on, the current is conducted through the vacuum arc. The LF current at t_3 is 7.1 kA and rising. One of the crucial parameters determining the probability of current interruption is the arcing duration of the VI – the duration between the moment of contact separation until the CZCs. It is related to the gap length between the contacts and hence, to the dielectric strength recovery of the VI, although the relationship to the latter is not linear [Sla07]. For the test result shown in Figure 9.5, the arcing duration until the 1st CZC is 2.9 ms even though the current interruption occurred at the 8th CZC after a total of 3.7 ms of arcing. This means the VI re-ignited during the first 7 CZCs. There are increasing (although not monotonic) re-ignition voltage spikes with alternating polarities seen in Figure 9.5a. This shows that the VI is indeed attempting to interrupt the current at each CZC.

The re-ignition voltage spikes observed at each CZCs are the ITIV due to the charge remaining on C_{inj} at the moments of CZCs. In reality, except at the 8th CZC, the re-ignitions occur before the entire ITIV appears across the VI. The prospective magnitude of the ITIV is not reached because the vacuum gap re-ignites already, due to insufficiently diffused arc residual plasma. The measured ITIV across the VI and the actual voltage across the capacitor are shown in Table 9.3 for comparison[9]. The increase in the re-ignition voltage at successive CZCs is due to the increased contact gap and, at the same time, the decreased $\mathrm{d}i/\mathrm{d}t$ at CZC. The latter is mainly caused by the decay of the HF current. A lower $\mathrm{d}i/\mathrm{d}t$ means the VI has more time to cool the arc. Otherwise at high $\mathrm{d}i/\mathrm{d}t$ a hot

---

[9]The ITIV higher than the actual voltage on C_{inj} at the 8th CZC is due to a superimposed HF oscillation caused by stray elements.

vacuum arc plasma still persists causing a re-ignition.

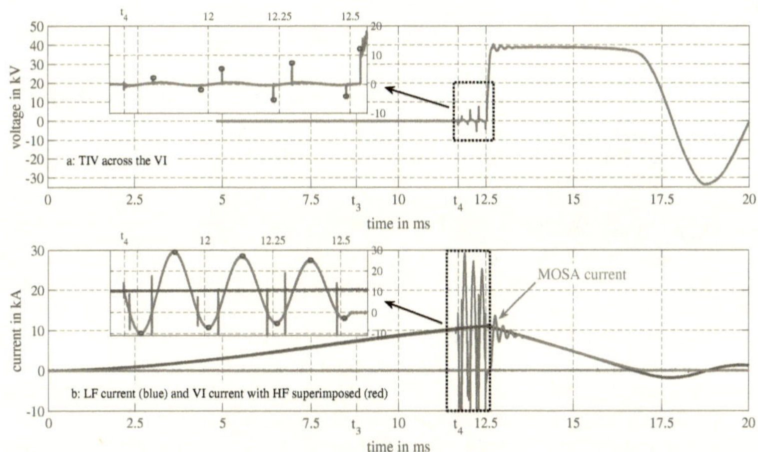

Figure 9.5.: Current interruption by the experimental DC CB – VI type A. ©[2020]IEEE.

From the zoomed portion of Figure 9.5b, it can be seen that, because of superposition of the LF current and the HF current, the current through the VI oscillates between high positive and low negative values (shown by blue circles at the local peaks). This results in major loop and minor loop currents before CZCs, and this has significant impact on the probability of a re-ignition at a CZC. There are assumedly two main causes for a re-ignition at a CZC: thermal and dielectric [BMS$^+$], the former being dominant during the major loop current flow. For example, the re-ignition at the 1st CZC seems to be entirely caused by the thermal effect as there is no observable re-ignition voltage. At the 2nd CZC, however, a re-ignition voltage of about 2.2 kV is observed. The crucial parameters near CZCs including the major and minor loop current durations are shown in Table 9.3.

In addition, closer scrutiny of all the test results shows that when a re-ignition occurs

---

[10] the peak value prior to a CZC
[11] arcing duration from contact separation until 1st CZC
[12] actual voltage across C_{inj} at corresponding CZC

Table 9.3.: Parameters at the CZCs during current interruption by a VI for the example case shown in Figure 9.5 ©[2020]IEEE.

CZC number	1	2	3	4	5	6	7	8
$\mathrm{d}i/\mathrm{d}t$ $(A/\mu s)$	454	440	368	356	290	286	209	206
peak current[10] (kA)	10.1	9.8	29.2	7.2	27.2	5.0	25.3	2.7
loop duration (μs)	2890[11]	83.7	170	75.2	180	64.1	191	51.3
re-ignition voltage (kV)	Negl.	2.2	1.6	5.5	5.26	7.3	4	12.5
prospective ITIV[12] (kV)	22.5	21.5	18.6	17.5	14.5	13.7	10.5	9.8

after a minor loop of the HF current (at even numbered CZCs), it occurs at a higher voltage compared to a re-ignition voltage following major loop current (at odd numbered CZCs). This suggests that re-ignitions after major loops of the HF current are aggravated by thermal effects. It also suggests that there are other crucial parameters that have an impact on the performance of the VI; the $\mathrm{d}i/\mathrm{d}t$ at CZCs and the local current peaks (of the HF current) just before CZCs. Studies show that re-ignitions at high $\mathrm{d}i/\mathrm{d}t$ are caused by post arc current [GBK72, GL72, STH$^+$84]. The post arc current is due to the metal arc plasma which remains in the vacuum gap immediately after current interruption.

From the moment of local current interruption at the 8th CZC onwards, the LF current is fully commutated to the injection branch of the DC CB, thus charging C_{inj} until the clamping voltage of the MOSA (the TIV) is reached. Henceforth, the MOSA maintains a more or less constant TIV voltage, for about 4.2 ms in Figure 9.5a, until the system current is suppressed. Even if there is no thermal energy being injected into the VI contact gap at this stage, the VI must withstand the TIV during the current suppression and subsequently the system voltage after the current suppression is over. The latter is not applied in this experimental investigation.

Nevertheless, it was observed on numerous occasions that the VI fails to sustain the TIV for sufficient duration even after sustaining the highest peak value. Henceforth, this kind of failure is referred to as a restrike[13]. Figure 9.6 shows a test result in which a restrike occurred during the current suppression period. The restrike happened about 1.2 ms after HF current interruption. Before the restrike, the LF current has been suppressed by about 3 kA from its peak value. After the restrike it starts to rise again until the VI clears the second time. This is a unique feature of an active current injection DC CB that a restrike may not necessarily lead to a complete failure to interrupt. The main impact of a restrike in this case is a longer total current breaking duration and an increased energy

---

[13]In this book, a re-ignition is a breakdown before the formation of full-TIV, whereas a restrike is dielectric breakdown after full-TIV is formed (after system current is commutated to EAB).

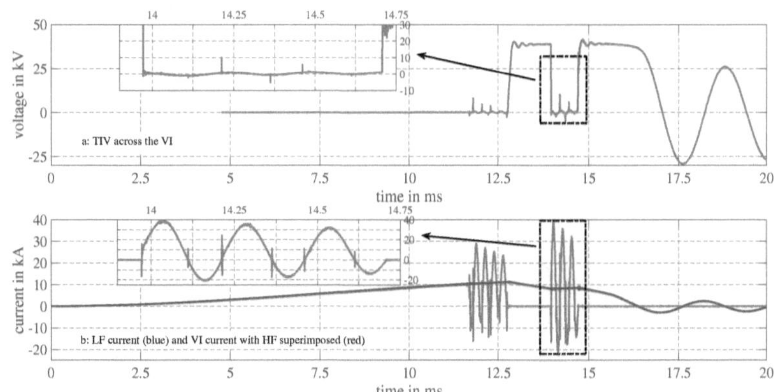

Figure 9.6.: Late restrike in the VI during the fault current suppression period. ©[2020]IEEE.

absorption in the MOSA. However, this is not always the case and the VI may not be able to clear after a restrike which was also observed on numerous occasions during the experiments.

9.3.2. Experimental results of VI type B

Similar experiments were performed using a VI of type B while keeping the rest of the circuit components and the test parameters as for the VI of type A. As mentioned in Table 9.1, VI of type B has AMF while VI of type A has TMF contact design. For VI of type B, there is a slight dispersion in the moment of contact separation and hence, the precise control of the arcing duration is difficult. The performance of the VI of type B regarding HF current interruption is, however, completely different than that of type A. For example, in about 75% of the experiments the VI of type B interrupted the current at the 1st CZC and, even at shorter arcing duration compared to the VI of type A. In about 16% of the tests this VI failed to interrupt at all. More than half of the unsuccessful current interruptions occurred during high-current tests (16 kA) on a single interrupter setup. Using a single interrupter setup, high-current test is performed 12 times, of which 6 times the VI failed to clear even if the LF arcing duration until the 1st CZC is prolonged to 3.5 ms. In the latter case 4 CZCs are created and the arcing duration until the last CZC is 3.8 ms.

Figure 9.7 shows low LF current interruption by a two interrupters setup of type B VIs. It can be seen that the most extreme attempt to clear is on the 1st CZC at which

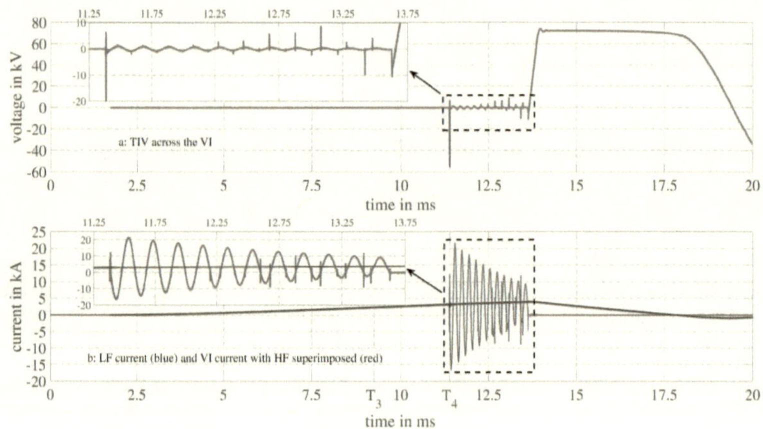

Figure 9.7.: Low current interruption by double-break interrupter of VI type B. ©[2020]IEEE.

re-ignition occurred at -50 kV ITIV. After the re-ignition there is no observable attempt to clear until after the 10th CZC. From the 10th CZC onward, there is an increasing re-ignition voltage until the HF current is interrupted at the 23rd CZC. The ITIV is the main cause of re-ignitions when interrupting low LF currents. This becomes severe for the two interrupters case especially when equal voltage grading across the VIs is not ensured.

Unlike the VI of type A where the probability of HF current interruption gradually increases at later CZCs, the major attempt to clear by the VI of type B is at the 1st CZC. This is observed in all the negative test results where the highest re-ignition voltage is seen at the 1st CZC suggesting that this type of VI has strong tendency to clear on the 1st CZC. In fact, there are attempts to clear on the later CZCs but the re-ignition voltages are much lower than at the 1st CZC, suggesting a lower HF current interruption capability after re-igniting at the 1st CZC. Of all the tests in which re-ignition occurred, only in less than 10% of the cases the VI of type B could interrupt at later CZCs, and there is no observable tendency to clear on even numbered CZCs (or after minor loop of the HF current) unlike the VI of type A.

Successful high LF current interruption is achieved when the LF arcing duration until the 1st CZC is prolonged to 3.6 ms. All the experiments with LF arcing duration longer than 3.6 ms resulted in successful interruption and, in all the cases, upon the 1st CZC. The main conclusion from the experimental results is that, for a given LF current, there is a minimum arcing duration that needs to be fulfilled before the CZC creation so that the VI will lead to successful interruption. Since the arc duration is directly linked to opening speed of the gap (the velocity with which the contacts move apart), the results of these experiments cannot confirm what is the main decisive factor: the LF arc duration or the gap-length at CZC.

9.3.3. Experimental results of VI type C

The third VI investigated is a VI of type C with rating and parameters as described in Table 9.1. Similar to the VI type B, VI type C has an AMF contact design. It also has a slight dispersion in the opening time, which in some cases, has resulted in unintentional sub-millisecond LF arcing duration before CZC creation.

For a single interrupter setup, only medium current tests have been performed at various arcing duration. For all the experiments, in which the LF arcing duration is in the range $3.8 - 4.5$ ms, the VI cleared on the 1st CZC and no restrike was observed. However, when the LF arcing duration is in the range 0.3-1.6 ms, restrikes were observed on a few occasions, which finally led to failed interruptions. Figure 9.8 shows a result, in which a restrike occurred in the VI type C. The VI failed to clear even though up to 18 CZCs are created after the restrike. In the test shown in Figure 9.8, the VI has been arcing for about 4.3 ms until the last CZC. This could have been sufficient for the VI type A to clear. The key conclusion here is that, this VI makes little attempt to interrupt the HF current after a re-ignition/restrike.

9.3.4. Summary of observations

The observations from the experimental results discussed in this section show that there is a clear difference in the HF current interruption performance of the three VIs used in the investigation. Very importantly, there is a clear difference between the VIs in the minimum LF arcing duration required for achieving successful current interruption. Table 9.4 summarizes statistical information of the overall experimental results of each type of VIs. It can be seen that the VI of type A show by far the most number of re-ignitions

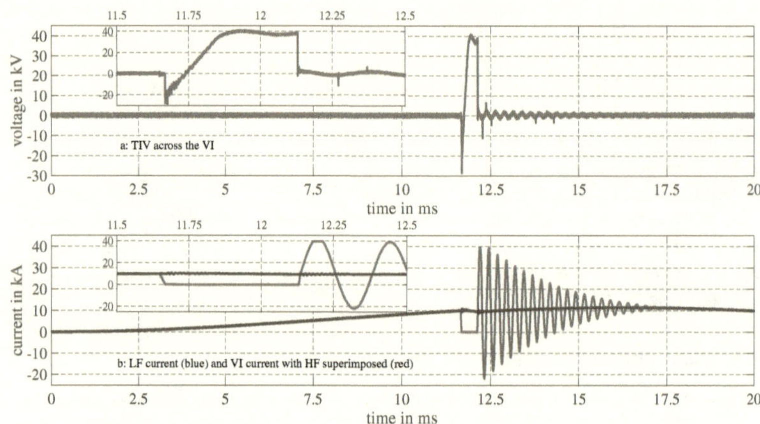

Figure 9.8.: Interruption failure after a restrike in VI type C. ©[2020]IEEE.

and restrikes with average of 5 re-ignitions per test. In fact only in 6.1% of the cases it cleared at the 1st CZC. Nevertheless, it does not exhibit the highest failure rate among the three types of VIs. It must be noted that restrikes in VI of type A do not lead to failed interruptions. Actually it clears the current in about 90% of the cases after a restrike. Type B VI has the least restrike rate but the highest failure rate. Most of the interruptions occurred at the 1st CZC. Unlike for the VI of type A, for the of VI type C all the restrikes led to failed current interruptions although the latter VI type has the highest rate of current interruption at the 1st CZC.

Table 9.4.: Summary of Test Results of the three VIs. ©[2019]IEEE.

VI type	#tests	#CZCs	1st CZC interruptions (%)	avg.# re-ignition	max.# re-ignition	restrike (%)	failure (%)
A	98	592	6.1	5.0	29	16.3	12.2
B	62	167	75.8	1.8	22	1.6	16.1
C	31	77	90.3	1.5	20	9.7	9.7

These differences could be attributed, for instance, to the design differences such as the magnetic arc control mechanisms. VI type A has TMF contacts whereas VI types B and C have AMF contacts as mentioned in Table 9.1. In general, further studies are needed.

For all the cases, as the LF current increases, the required minimum arcing duration until CZC creation tends to increase. Moreover, there is maximum LF current that can be interrupted by any VI within the relatively short arcing duration required in the DC CBs. The latter could be related with an anode spot formation at high current. As current surpasses a certain threshold, the arc constricts towards the anode because of electromagnetic force, thus creating an anode spot [Lan08]. The anode spot takes more time to cool. When an anode spot leads to local melting of anode material of considerable size (in the millimeter range), solidification and emission of molten material continues after current zero, hence impacting recovery of the gap. After current zero the anode becomes a new cathode. Thus, the anode spot not only enhances contact wear but also increases the probability of re-ignition. In AC applications, arc control in VCBs is intended to avoid anode spots, and it has been demonstrated that this method can increase the AC current interruption capability to 63 kA and above. However, this is always under a well developed gap-length. In the present application, the behavior of the arc plasma under short LF arc duration and hence, minimum gap-length is essential. In the literature, no information is available how the arc control (AMF, TMF) impact the arc plasma and its capability to interrupt HF current and/or withstand capability of short-time DC TIV. In addition, the ITIV is a phenomenon not known in AC applications.

An important conclusion is that a test must ensure sufficient duration of TIV to verify the dielectric recovery of the VI(s) for such application. This has significant impact in the energy absorption of the DC CB and is discussed in the next section. In addition, it is shown that re-ignitions are common not only during high-current interruptions but also during low current interruptions – showing the importance of low current interruption in the test requirement specification. In general, further studies are needed to investigate whether re-ignitions and restrikes are acceptable to the HVDC system although the current is cleared after all.

9.4. Experimental Results: Thermal Stresses on the Metal Oxide Surge Arrester (MOSA)

The other key constituent of all HVDC CBs, which is subjected to unique stresses during DC breaking, is the energy absorbing component, i.e., the MOSA. In HVDC CBs, the MOSA serves two main functions: clamp and maintain the TIV to a desired level, and absorb the system energy during current suppression. The MOSA must be designed to meet the requirements of both functions simultaneously. The desired level of the TIV must be sufficiently higher than the system operation voltage while the energy absorption is dependent on the system as well as on the circuit breaker parameters [BPS18b]. A TIV about 50% higher than the system voltage is considered as a reasonable value for system insulation coordination as well as current interruption duration [Cig17a].

Conventionally, MOSA is used for overvoltage protection in power systems. A few major differences between the use of MOSA for overvoltage protection (both in AC and DC applications) and for HVDC CB applications are highlighted in [HBHS].

9.4.1. Design of a MOSA for HVDC CB application

Commercially available MO varistors of different makes have been evaluated for HVDC CB application in [BH19]. Extensive endurance tests have been performed where the MO varistors are subjected to >10000 impulses at energy per volume in the range of 60-100 J/cm^3. It is important to learn about the aging phenomena after such large number of impulses, especially with DC voltage application following each impulse [HBHS]. Since there is no international standard for HVDC CB application, the number of impulses in the endurance test are borrowed from the AC CB standard whereas the pass/fail criteria are according to the surge arrester standards. However, such a large number of switching operations (involving energy absorption) is not realistic in HVDC systems. Realistic parameters of the applied impulse current shall also be utilized in such investigations – typically several milliseconds duration is expected in actual applications.

It is reported in the literature that the energy handling capability of an MO varistor depends on several factors such as the diameter and volume, impulse duration, current density, etc., [Eda, RKE$^+$97, Cigb, Cigc]. Moreover, different makes of MO varistors exhibit considerably varying energy absorption capabilities [TH14, Tuc, Cigb, Cigc]. Nevertheless, the optimum energy absorption per volume, considering a large number of MO varistors in a MOSA module of an HVDC CB, needs further investigation.

When designing a MOSA module for HVDC CB application, the desired TIV determines the residual voltage (and thus the height) of the active part whereas the expected energy in the system determines the total volume. Since a large amount of energy is absorbed during DC fault current interruption, several parallel columns of MOSA are required to cope with the volumetric requirement. However, this requires a column matching procedure: a crucial design step when constructing a multi-column MOSA. This column matching procedure is necessary to ensure equal current sharing during the fault current suppression, which otherwise would lead to unequal energy distribution and hence, possible thermal overloading of one or more columns.

Note that typically after manufacturing, all MO varistors are screened by applying 8/20 μs impulses to check the U-I characteristics, specifically by measuring the residual voltage at 10 kA impulse current. Even after this, not all the MO varistors have identical U-I characteristics and hence, the MOSA columns built from the same batch of MO varistors do not necessarily have matched U-I characteristics. This is due to inherent imperfections in the manufacturing process that do not always lead to MO varistors with an identical distribution of microscopic ZnO grains. The voltage of the MO varistor is determined by the number of ZnO grain boundaries conducting along the current path. Thus, it is difficult to ensure a homogeneous distribution of grain sizes and/or impurities along all current paths even within the same MO varistor batch. Current initially flows in the path that results in the fewer number of grain boundaries until it is distributed across the entire cross-section. This is what results in localized current conduction especially at low current densities [BCM99].

The issue of MOSA column matching has been identified in the early developments of HVDC CBs [YTI$^+$82]. Different column matching techniques exist. During the column matching procedure, current impulses are applied (10-20 times) successively to the parallel arrangement of MOSA columns – see an example of impulse measurement through six columns in Figure 9.9b. One among these columns serves as a reference column as shown in the Figure 9.9a. The arithmetic mean of the peak values of the current through each column is computed, and compared against the arithmetic mean of the peak values of the current measured through the reference column. Columns with current measurements within acceptable margin, for example ±3% from the reference column current, are accepted as matched[14]. The procedure is repeated until the required number of columns are obtained.

Another similar routine procedure of column matching is by applying an impulse current of around 500 A per column [ABB]. In this case, by measuring an impulse current through each column, the deviation from the arithmetic mean is calculated after 20 impulses. The selection criterion is that those MO arrester columns with a maximum deviation of $+7\%$ from the arithmetic mean are accepted as matched[15]. In fact the columns are constructed from initially pre-selected MO varistors. The MO varistors are pre-selected by applying impulse current of 500 A and 10 kA of 8/20 μs.

9.4.2. Specification of a MOSA module for an experimental DC CB

The specification of one module of MOSA is determined based on the rating of the experimental DC CB, which is specified as follows, see Section 9.2.2:

---

[14] ±3% in current sharing typically means <<1% in residual voltage difference, due to the extremely high degree of non-linearity in the considered current range of only a few hundred amperes per column in many cases.

[15] in this case there is no reference column for comparison

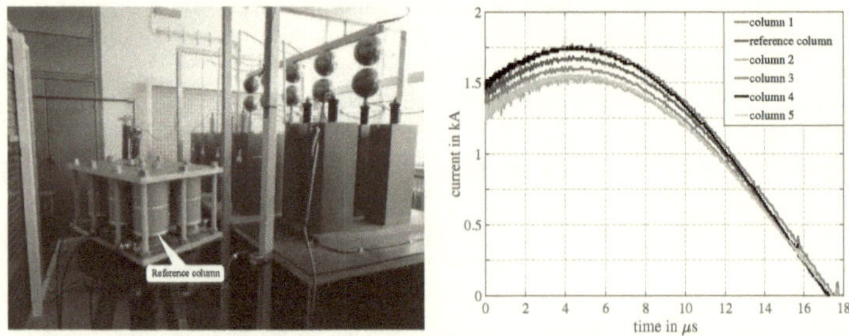

(a) An example of MOSA column matching procedure test setup (b) An example of impulse current measurements

Figure 9.9.: MOSA column matching for high energy absorption. (courtesy: TU Darmstadt)

- Transient Interruption Voltage (TIV) – 40 kV

- Rated interruption current – 16 kA

- Rated energy – 2 MJ

Thus, based on the above specification, the height and the volume of a MOSA module are determined using electric field, current density and energy per volume relationships. Then, a proper MO varistor is chosen (preferably the largest diameter that can be manufactured in order to reduce the number of parallel columns although the specific energy absorption capability of large diameter MO varistors is often slightly lower than that of smaller diameter MO varistors [TBHG15]). A MO varistor with diameter 99±1 mm and height 21.4±0.6 mm with a nominal residual voltage of 7.4 kV at 10 kA discharge current is selected. It is reported in the literature that an MO varistor can handle over 400 J/cm^3, [RKE$^+$97, BCM99]. To be on the safe side, a 200 J/cm^3 is assumed [Cigb] (which is also a typical design energy of many MO arrester manufacturers) and based on this, the volume of MOSA required for 2 MJ energy absorption is 10000 cm^3. This results in 60 samples of MO varistors of the selected dimension. To meet the TIV specified above, 6 MO varistors need to be put in series to form one column – see Figure 9.10. This results in 10 parallel columns per MOSA module. Additionally, 2 columns are included to reduce the energy per volume to a safer value (170 J/cm^3) resulting in 12 parallel column per module[16]. Besides, since it was intended for indoor experimentation, MOSA columns with bare varistors are used without any kind of housing as shown in Figure 9.11a. Using the above procedure and information, 9 MOSA modules with the same specification are constructed for investigation in an experimental DC CB. The specified MOSA column current sharing criterion was ±5% although later it was reduced to ±3%.

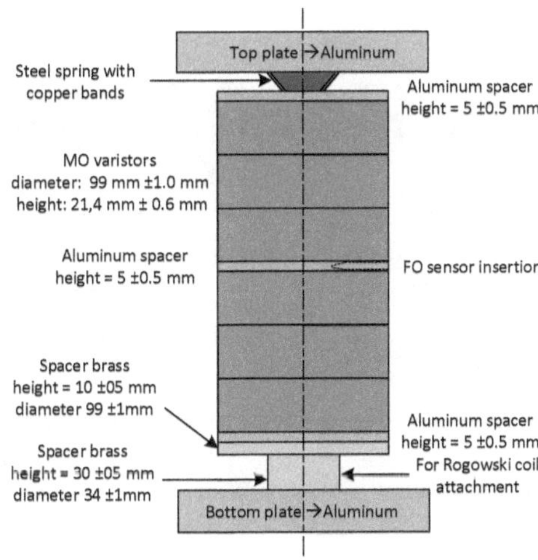

Figure 9.10.: Details of MOSA column design – showing access for temperature and current measurements

9.4.3. Discussion of MOSA stresses

The thermal energy handling performance of the MOSA modules is investigated under real-power test conditions where different energy levels are successively injected during current suppression in the experimental DC CB after the HF current interruption. During the experiments, currents through 8 columns are measured using Rogowski current probes

---

[16]However, at a rated current interruption, the resulting residual voltage of the MOSA module reduces by more than 15% compared to that of a single column due to the reduced current density

(CWT30LFxB/4/300), and the temperature of the corresponding columns is measured
using 8 Qualitrol optical temperature sensors (T2S-20-01-1, measurement range -80-250 °C,
accuracy ±1 °C and response time 250 ms) with an Omniflex-2 (OFX2-CHA3U) signal
conditioning system, see Figure 9.11a. As can be seen from the figure the fiber optic (FO)
temperature sensors are mounted inside an aluminum plate placed between MO varistors
at the midpoints of the MOSA columns. The temperature sensor does not directly touch
the surface of the MO varistor, rather through a thin plate of aluminum and thermal
compound filled sensor aperture as shown in Figure 9.10.

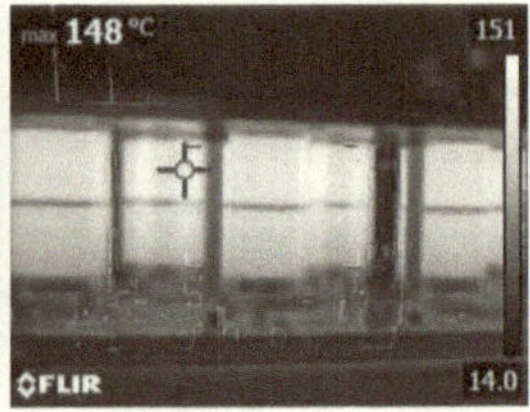

(a) Experimental setup showing actual current and temperature (b) Thermal image of a MOSA
    measurements of 8 columns of a MOSA module   module

Figure 9.11.: Details of MOSA column temperature and current monitoring. ©[2020]IEEE.

Moreover, the MOSA columns' surface temperature is monitored using a thermal
imaging camera (FLIR E85) as shown in Figure 9.11b. The MOSA module columns are
arranged in such a way that each column can be "seen" by the thermal imaging camera
from the front/rear side. Also the voltage across the MOSA module is measured using a
NorthStar VD150 voltage divider as can be seen in the test setup shown in Figure 9.2.

Figure 9.12 shows the MOSA temperature measurements using FO temperature sensors
during 18 consecutive current interruptions. The same test (approximately equal energy
of 900 kJ) is repeated until the temperature of the MOSA reaches about 170 °C. Initially,
the MOSA column temperature rises (on average) by about 25.7 °C, and the temperature
of the columns is similar within a few degrees showing uniform energy sharing.

However, the temperature differences between the columns increase as the MOSA is
heated by successive energy absorption before cooling down. This is partly related to
the differences in cooling of the MOSA columns due to their physical arrangement. The
columns located in the middle have less heat dissipation compared to the columns located
at the edges. Nevertheless, the increase in the temperature per energy injection is slightly
reduced as the MOSA temperature rises. This can be seen from the zoomed sections of

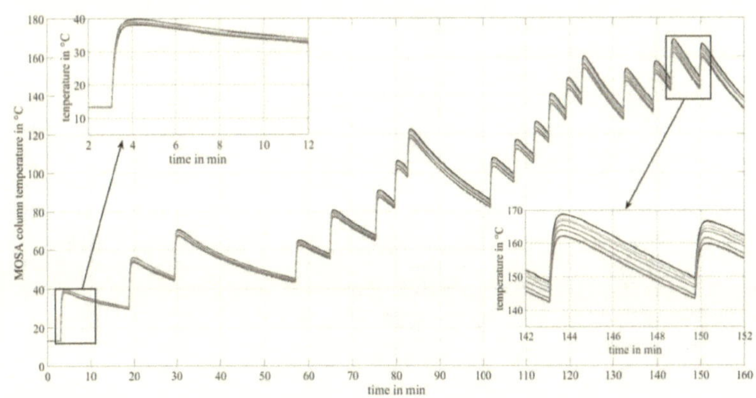

Figure 9.12.: MOSA columns temperature measurement over successive current interruptions – # 1, 17 and 18 are magnified. ©[2020]IEEE.

Figure 9.12 where the temperature rise during the last test is about 20.1 °C, see Table 9.5. The table shows more than 5 °C difference in the temperature rise compared to that of the first test for roughly the same amount of energy injected. The reduction in the temperature rise is due to the increase of the heat capacity with temperature [Lat, Cigb], in this case from 2.85 kJ/°C to 3.53 kJ/°C.

Table 9.5.: Temperature rise versus energy absorption of MOSA columns at different initial temperatures: Test #1 at ambient and Test #17 at 145 °C ©[2020]IEEE.

Column #	Test #1		Test #17	
	Energy (kJ)	Temp. (°C)	Energy (kJ)	Temp. (°C)
1	71.14	25.7	69.32	20.6
2	68.92	25.0	69.98	19.8
3	70.91	25.2	68.43	20.0
4	75.61	25.5	72.81	19.9
5	76.11	26.6	71.50	20.3
6	75.30	26.4	73.08	20.3
7	73.56	25.2	73.08	20.0
average	73.07	25.66	71.17	20.13

In addition to the total current through the MOSA module, the currents through 8
columns of the MOSA are also measured to monitor current sharing. An example of
measurements of current through MOSA module columns are shown in Figure 9.13a. It
can be seen that the currents through the MOSA columns are not equal. In fact, this is
the current measurement obtained when two MOSA modules are connected in parallel. It
can be seen that column 2 conducts less than 66% of the current in column 6 and hence
absorbs similar proportion of energy as shown in the bottom graph. This is likely due to
the difference in stress history of the two modules and it shows that care must be taken
when combining MOSA columns or modules with different stress history/aging. Even if
the MOSA modules are adequately matched, the difference in stress history results in
changes in the residual voltage one compared to the other. For example, 0.1% change in
residual voltage may easily cause more than 10% change in the current sharing.

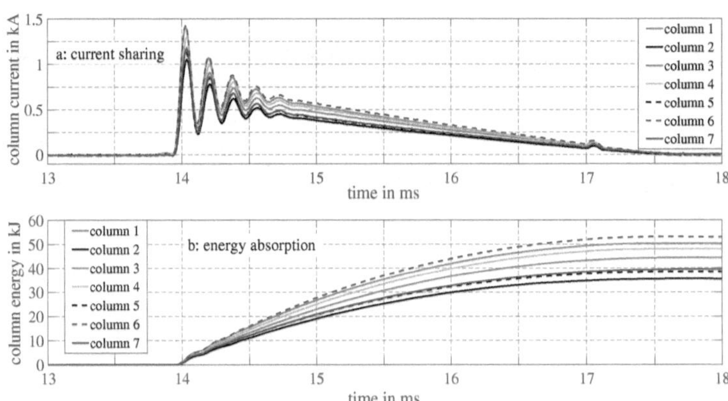

Figure 9.13.: Current per column and energy absorption per column of the MOSA module
during a typical test.

In order to investigate the impact of column mismatch, one of the 9 MOSA modules
is designed to have a column (column 1) that deviates in current sharing by 10% from
the reference column. Hence, this column is intentionally designed to draw 10% more
current than the reference column. Successive current interruptions have been performed
using this module. Figure 9.14 shows the temperature measurement of 8 columns (one of

which is column 1) over 7 successive current interruptions. About 1 MJ energy injection is targeted at each test. The impact of the increased column mismatch can be clearly seen from this figure where the red trace shows the temperature measurement of the mismatched column 1. As can be expected the temperature of the mismatched column 1 is clearly higher than the temperature of the other columns – about 10% higher than the lowest temperature of the other monitored columns. The main lesson here is that at the energy absorption per volume of <100 J/cm^3, the 10% deviation is not so significant to cause any concern. In fact for some similar applications such as the MOSA for the protection of series compensation capacitor banks, MO varistor columns are designed to withstand current deviations of up to 10% [ABB].

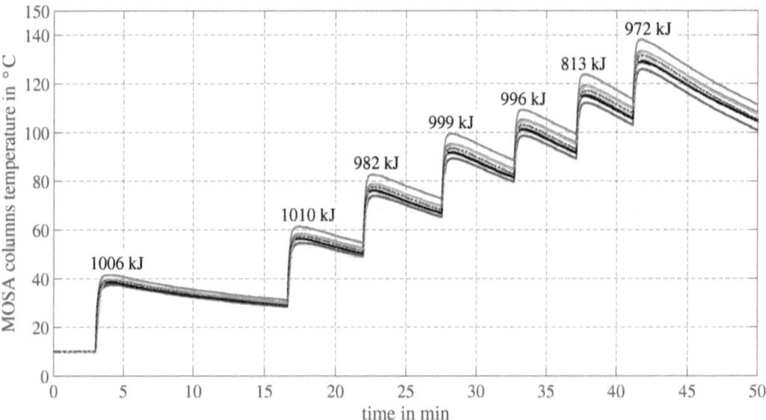

Figure 9.14.: MOSA module column temperature measurement over 7 successive current interruptions – One of the columns is designed to draw 10% more current than the others. The number at each temperature step indicates the total energy absorbed by the module.

Note that during the successive current interruption tests discussed above, the maximum energy absorbed by a MOSA module per a test is about 1.0 MJ which is less than half the design value of the energy absorption. This is limited intentionally in order not to overheat the MOSA while performing as many experiments as possible within a short period of time (as part of VI investigation).

Later, a few experiments are performed on the two interrupters DC CB while injecting 2.6 MJ energy per MOSA module. This results in energy per volume of over 200 J/cm^3, which is generally a value considered as acceptable to MO varistors. The result of one of these tests is shown in Figure 9.15 where the electrical stresses on the MOSA, namely, current and voltage[17] are shown. The MOSA is conducting current for nearly 10 ms, which is an extremely long duration compared to the lightning/switching impulse duration in the conventional overvoltage protection application in power systems. The test shown in Figure 9.15 is repeated 4 times in quick succession to investigate the thermal performance limit of the MOSA modules under the specified energy absorption.

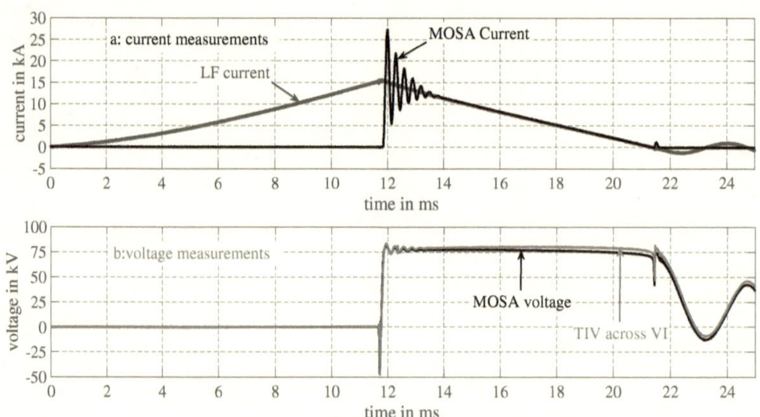

Figure 9.15.: Current through and voltage across MOSA, and TIV across the VI during current interruption by experimental DC CB – high-energy absorption. ©[2019]INMR

The temperature measurements of two columns (one column from each MOSA module) are plotted in Figure 9.16. The temperature rise and the energy absorbed at each test is labeled in the figure. In this case a temperature rise by 72 °C is observed during the first

---

[17]The TIV across the VIs is slightly higher than the voltage across MOSA due to some stray inductance in the loop.

test where 5.2 MJ energy is injected. In the next test a temperature rise of 68.6 °C is observed for nearly the same amount of energy whereas in the third test a temperature rise of 66.3 °C is measured. The reduction in temperature rise per energy mirrors the increase of the heat capacity of the MO varistors with temperature. By the third test the MOSA temperature exceeds 200 °C.

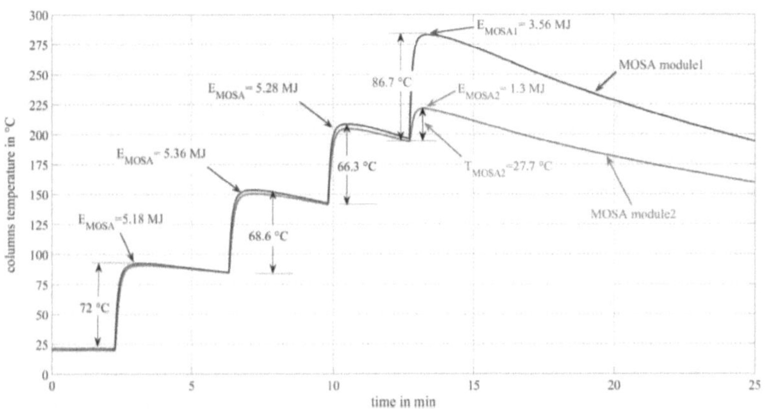

Figure 9.16.: Temperature measurement during successive high energy tests until destruction. ©[2020]IEEE.

The fourth test is performed while the MOSA temperature is around 200 °C. Actually, the MOSA failed in the fourth test during which a few MO varistors cracked and some were punctured. The failure of the MO varistors led to a flashover between the terminals of the MOSA module. Moreover, it is interesting to observe that the two MOSA modules did not fail at the same moment. Rather, the failure of MOSA module 2 caused the failure of the MOSA module 1. This is because when one MOSA module fails, the remaining MOSA module has to deal with the remaining energy in the circuit. The blue trace in Figure 9.16 shows the temperature of MOSA module 1 while the red trace shows the temperature of MOSA module 2. During the final test, the temperature of MOSA module 2 increased only by 27.7 °C whereas the temperature of MOSA module 1 increased by 87.7 °C. Considering the additional energy and the temperature rise, it is likely that the MOSA module 2 is pre-damaged before this test although there was no visual indication

as such. After MOSA module 2 failed, the MOSA module 1 kept conducting normally for
about 5.5 ms after which it also failed. The MOSA module 2 absorbed 3.56 MJ (40%
more than its (recommended) design value) of energy before it failed itself.

The actual current and voltage measurements during the destructive tests are shown in
Figure 9.17. Measurements of the LF current, current through the MOSA modules and
the TIV are shown. It can be seen that both MOSA modules initially conduct normally
for about 2.2 ms until MOSA module 2 fails leading to a flashover between its terminals.
This results in a TIV drop by 50% as can be seen in the TIV plot (part b) of the figure.
Since the TIV is more or less equal to the source voltage at this moment, the system
current is no longer suppressed, instead it is limited more or less to a constant value
of about 10.5 kA for about 5.5 ms after the MOSA module 2 fails. The long duration
current conduction of MOSA module 1 has led to localized current conduction which
ultimately results in localized overheating. The latter resulted in punctures of the MO
varistors as can be seen in Figure 9.18d.

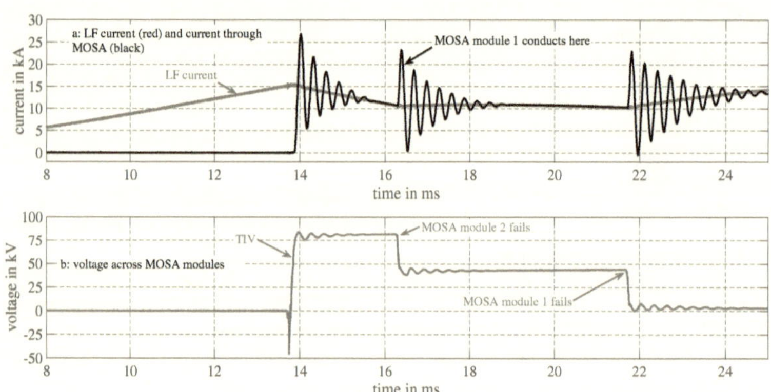

Figure 9.17.: Current and voltage measurement during MOSA destruction test.
©[2020]IEEE.

Not only the instants of failure but also the failure modes of the two MOSA modules
are different – see Figure 9.18. The MO varistors in Figure 9.18a and 9.18b belong to
MOSA module 2 where one of the varistors cracked before flashover occurred between
its terminals – this can be verified from the voltage measurement when MOSA 2 fails

in Figure 9.17. After the test, all the MO varistors in the MOSA modules are visually inspected.

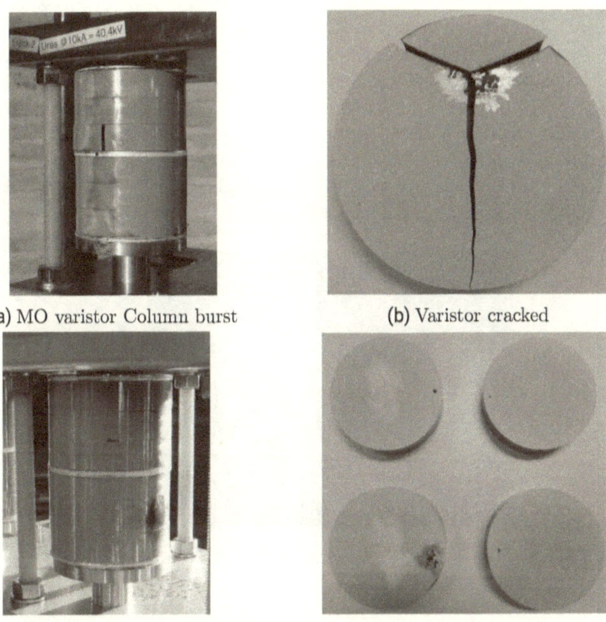

(a) MO varistor Column burst   (b) Varistor cracked

(c) Surface arc tracking   (d) Varistor punctured

Figure 9.18.: Failure modes of MO varistors after destructive tests. ©[2020]IEEE.

Of all the 72 MO varistors in the MOSA module 2, only one MO varistor (shown in 9.18b) of one column (shown in 9.18a) cracked while others are visually clean and do not show any trace of damage. Cracking occurs due to thermo-mechanical stress caused by nonuniform heating of MO varistor body over a short duration at high temperature gradient [Hin11, BCM99]. This shows a failure of a single MO varistor is sufficient to cause a failure of the entire MOSA module. Figure 9.18c and 9.18d belong to MOSA module 1 where punctures of varying diameters are observed in several varistors of many columns. The failure of MOSA module 1 also led to flashover between its terminals as evidenced from current and voltage measurements in Figure 9.17. In MOSA module 1 many varistors are punctured and in some cases tracking on the coating of MO varistors is observed. Punctures occur due to localized heating when MO varistors conduct current for long duration which is the case for MOSA module 1.

A poorly matched set of MOSA columns could lead to a successive failure of MO varistors in a column. When not sufficiently matched, one or more columns of a MOSA module draws more current (and hence more energy) compared to other columns. Figure 9.19 depicts a thermal image of a MOSA module, in which one of the columns is heated compared to the remaining columns. In fact this is an observation at a test with much lower energy absorption. This effect could cause a failure if a nominal energy of about 200 J/cm^3 had been injected.

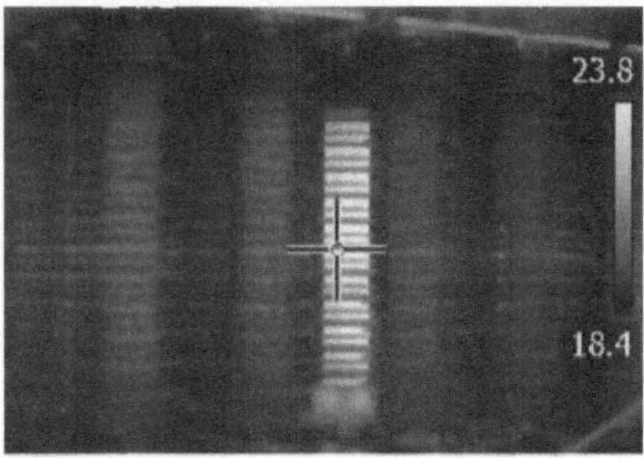

Figure 9.19.: Impact of unequal energy sharing – resulting in unequal heating of MOSA columns

An oscillogram of a typical failure of MOSA, which occurs as a result of poorly matched columns, is shown in Figure 9.20. In this case three modules of MOSA banks are used in cascade in parallel to three series connected VIs. The waveforms of current interruption during successful operation[18] (the green traces) are superimposed together with the prospective current (dashed black trace) to provide reference for comparison. The TIV and the interrupted current during a test, in which MO varistors fail are shown by blue and red solid traces, respectively. Current suppression started normal (as expected) for the first 500 μs until an MO varistor fails in one of the parallel columns. This can be seen from the steep drop on the TIV (seen at t_1 in the zoomed plot). The voltage difference at this point corresponds to a voltage of one MO varistor. A failure of this varistor led to

---

[18]Slightly higher current is interrupted due to delayed trip command in this case.

short-circuit while the remaining series connected MO varistors conduct normally. Due to the reduced overall voltage of this column, almost the entire suppression current flows through this column. This column could sustain the voltage and current for about 400 μs until the next MO varistor fails. This can be seen at t_2 in the zoomed part of Figure 9.20. Then, immediately the next 2 MO varistors (at t_3) and the last two failed at t_4 in quick succession as can be seen in the zoomed section of the figure. Even if one of the three MOSA modules failed, the short-circuit current has been suppressed by the remaining MOSA modules.

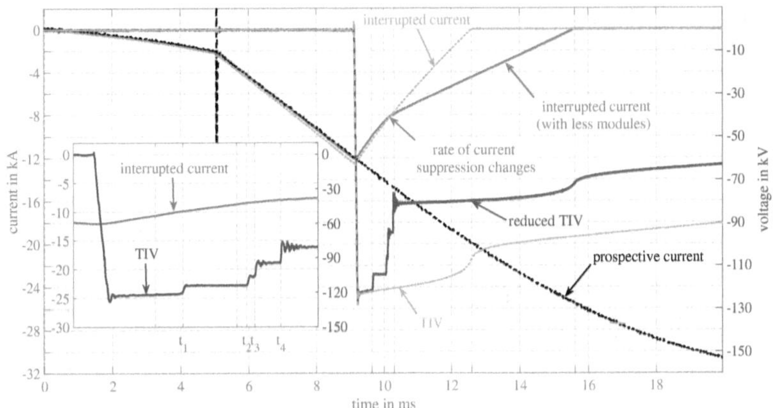

Figure 9.20.: Successive failure of MO varistors in a column of a MOSA module during current suppression

From the interrupted current plot, it can be seen that the rate of current suppression reduces during the successive MO varistor failures. This resulted in a prolonged current suppression duration – nearly twice the expected current suppression duration as can be observed in comparison with the green dashed trace. The prolonged current suppression duration in turn results in increased energy absorption in the remaining MOSA modules. In this case, in total 3 MJ[19] energy is absorbed by the two MOSA modules as opposed to

[19]Much higher energy could be absorbed if a source voltage equivalent to the rated voltage (80 kV) of the CB had been used in the circuit. In this case a source voltage with peak value of 55 kV is used.

2.4 MJ expected energy absorption of the three MOSA modules. A critical point is that this energy is absorbed by two MOSA modules, which shows the need for considering sufficient margin in the energy absorption when designing MOSA modules for HVDC CB. An important lesson from this is the criticality of losing only a single MO varistor in a multi-column structure. This requires due attention when designing MOSA because a large number of MO varistors are needed for this application, and the probability of losing one MO varistor significantly increases compared to failure probability of a single MO varistor investigated in the literature.

Sufficient energy absorption margin is necessary also because in some cases higher current might be interrupted over a longer current suppression period than normally expected due to multiple re-ignitions and restrikes. Figure 9.21 shows a test result in which a VI re-ignited and restruck several times during current interruption. As a result of several re-ignitions the short-circuit current could rise to 18.2 kA instead of expected 16 kA.

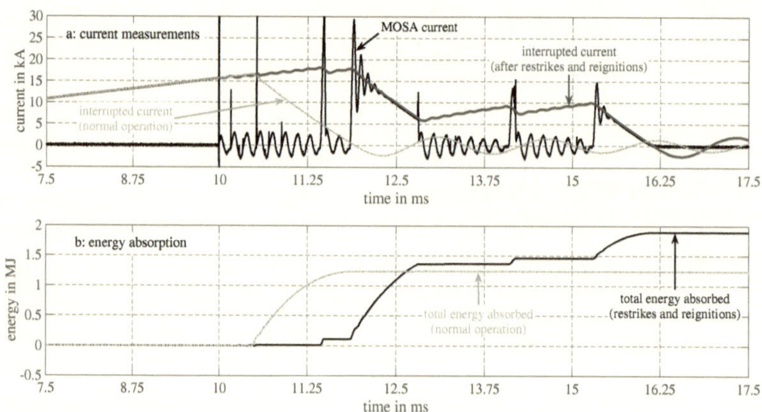

Figure 9.21.: Multiple restrikes leading to accumulative energy absorption in the MOSA

The green traces in the figure show another test result, in which a breaker could interrupt the short-circuit current as expected. The current through the MOSA during the current interruption process is shown by the black trace. In this case the HVDC CB could clear the short-circuit current after all, although at remarkably prolonged current

suppression (> 4 ms prolongation) and much higher energy (45% higher) than expected
as shown in the Figure 9.21b. Also in this case a much lower energy than the rated energy
of the HVDC CB is targeted, thus, even with the amount of additional energy due to the
re-ignitions and restrikes the total energy remains far below the energy that the MOSA
branch can handle. In this case, a source voltage of around 20 kV is used to supply the
test circuit as opposed to the rated 80 kV voltage. However, the main lesson in this case
is the need to have sufficient margin in the energy absorption capability of the MOSA
under rated short-circuit current interruption conditions.

9.5. Summary and Conclusion

The chapter presents the results of the experimental investigation of two major components
of HVDC CBs: namely, VI and MOSA. Three different designs of VIs have demonstrated
completely different HF current interruption performance when interrupting DC short-
circuit currents. This shows that VIs can be optimized for HVDC CB application. Factors
determining the performance of VIs are the minimum arcing duration (contact gap
distance), maximum interruption current, the rate-of-change of current at CZC, the peak
value of the TIV, the rate-of-change of TIV, and the duration of the TIV. Moreover,
re-ignitions and restrikes are very common which entails that the test methods need to
take into account adequate voltage stress across the insulation gap (vacuum, SF_6, air)
during the entire interruption process. The importance of testing for sufficient duration
of TIV, and the criticality of low current interruption is verified.

The prominent features of MOSA for HVDC CB application has also been investigated.
A robust design including adequate matching of columns to ensure equal current sharing
between all columns is essential. A properly designed MOSA can function adequately even
at elevated temperature up to 200 °C while handling energy per volume up to 200 J/cm^3.
This shows that, if this energy absorption requirement is fulfilled, it may not be necessary
to double the volume of the MOSA for re-closing purposes for instance. Applying energy
at temperatures higher than 200 °C may result in a damage of an MO varistor(s) which
leads to overall failure of the HVDC CB. Different failure modes, namely, varistor cracks
and punctures are observed. Nevertheless, a poorly designed MOSA module (for example,
with mismatched columns) can drastically worsen the overall performance of an HVDC
CB. Moreover, in order to reduce risk of damages or even explosions following failures of
MO varistors, especially during high-energy tests, it is advisable to monitor the individual
column temperature and/or current sharing. The methods used in this investigation
include the use of thermal camera, individual column current measurement or actual
column temperature monitoring through FO sensors, which are found to be effective.

10. Summary, Conclusion and Recommendation for Future Studies

10.1. Summary and Conclusion

The study in the book provides insight in both HVDC CBs and the MTDC grids in which they are operated. In the absence of any international standards and/or recommendations for testing of HVDC CBs, it sheds lights on the critical factors that need to be considered when developing test requirements. It provides crucial inputs to the ongoing (at the moment of writing) international standardization activities.

First, a fault condition in a conceptual MTDC grid and the main contributors to the fault current at different stages the fault current development are investigated through system simulation. Following this, the critical stresses on HVDC CBs along with the important stages of the fault current interruption process are identified. These stresses are translated into test requirements for HVDC CBs. Also, the requirements for a test circuit are specified based on the identified st resses. A test circuit capable of reproducing all the necessary stresses in each stage of the fault current interruption process is designed, implemented and demonstrated with testing of the actual industrial prototypes of HVDC CBs at their final stages of product development.

In an MTDC grid, a fault current develops in several stages with various system components and parameters affecting the rate-of-rise and m agnitude. In general, three temporal stages of fault current development are identified o nce t he t raveling waves emanating from a fault location arrive at an HVDC station:

1. Converter SM capacitor discharge stage

2. Arm current decay stage

3. AC grid in-feed stage

The duration of each stage, and the rate-of-rise of the fault current in each stage depend on the system design parameters including the size of the SM capacitors, the DC current limiting reactor, the arm reactors, the number of connected lines at the DC busbar, the short-circuit capacity of the AC grid as well as the converter fault-ride-through capability.

From an HVDC CB perspective, the worst-case situation is when an MTDC grid fault results in a high short-circuit current under stiff DC voltage regulation. This corresponds to a fault near a converter terminal and when an HVDC CB clears the fault while the converters continue their controlled operation (without blocking).

Four critical stages of the DC fault current interruption process have been defined: (1) internal current commutation, (2) TIV generation and maintenance, (3) energy absorption and (4) DC recovery voltage withstand. It is found that it is essential to verify the correct performance of HVDC CBs during each of these stages. The key design and performance parameters of an HVDC CB at each stage have been identified.

A test program covering the current interruption requirements of an HVDC CB while representing the worst-case stresses in terms of current, energy and dynamic TIV is defined:

- TF100 – 100% rated peak fault (maximum) current interruption at rated energy

- TC100 – 100% rated continuous current current interruption

- TC10 – 10% rated continuous current current interruption

In case of a limitation in energy absorption either due to the test circuit or the DUT, a multi-part testing procedure has been proposed as an alternative: TF100* for the 100% rated peak fault (maximum) current interruption and TDT for the rated TIV (fault current suppression) duration, both at reduced energy absorption.

The requirements of a test circuit capable of reproducing the essential stresses is defined. In each test duty, a test circuit shall provide current, voltage and energy to verify the correct functioning and the performance of HVDC CBs at each stage of the current interruption process preferably in a single test. An adequate and complete test circuit shall supply to the HVDC CB,

- The short-circuit current with adequate rate-of-rise while considering the worst-case fault neutralization time. The short-circuit current shall rise to at least the peak fault current interruption capability (TF100) of the HVDC CB.

- Adequate energy as in service. The energy includes not only the magnetic energy stored in the system inductance but also the electrical energy injected by the system. For this a test circuit shall maintain sufficiently high source voltage not only during the fault current neutralization period but also during the fault current suppression period.

- DC recovery voltage equivalent to the system voltage after current suppression.

A realistic test method that utilizes test circuit components available at most short-circuit test facilities is proposed and demonstrated. The test method is based on the use of AC short-circuit generators operated at low power frequencies, e.g., at 16 2/3 Hz. The test method maintains a quasi-DC supply voltage while providing short-circuit current with sufficient rate-of-rise and magnitude. It is verified that this test method can supply the required current, energy and voltage stresses during the current interruption process to the state-of-the-art HVDC CBs.

A pragmatic approach of testing HVDC CBs has been developed, realized and verified at a high-power laboratory using three industrial prototypes in the last phase of product development as test and demonstration objects. Many additional features and sub-circuits needed to ensure the correct and safe operation of the test circuit while applying complete stresses in a single test have been implemented and demonstrated. The performance of the following HVDC CBs with voltage and current ratings, a specified energy absorption and breaker operation times have been demonstrated,

- Active current injection – direct discharge of pre-charged capacitor 160/200 kV, 16/20 kA, 5 MJ, 7ms

- Hybrid – original topology 350 kV, 20 kA, 10 MJ, 3 ms

- VSC assisted resonant current (VARC) 80 kV, 12/15 kA, 3 MJ, <2 ms.

The limitations of the method and test circuit have also been identified, and methods to address some of these limitations are proposed. These include limitations to supply the required current, voltage and energy stresses in a single test to the HVDC CBs rated above 500 kV. The other critical challenge is the transformer saturation, as a result of which the required test parameters cannot be achieved. Practical methods to mitigate these challenges have been proposed and demonstrated.

Two major components of HVDC CBs: namely, VIs and MOSA have been investigated in an experimental DC CB setup at medium voltage level. Three different designs of VIs have demonstrated completely different performance when interrupting DC short-circuit currents. This shows that VIs can be optimized for HVDC CB application. Experimental results justify the importance of testing the realistic TIV duration.

The key feature that becomes prominent in the use of large volume MOSA for HVDC CB application is the column matching to ensure equal current sharing and hence, the energy absorption. A properly designed MOSA can function adequately even at elevated temperature up to 200 $°C$. On the other hand, a poorly designed MOSA module(s)can drastically worsen the overall performance of an HVDC CB.

10.2. Recommendation for Future Studies

- The interaction of an HVDC CB with the HVDC system needs further investigation. Insulation coordination considering the superposition of traveling waves and the TIV produced by the HVDC CB should be studied. The overvoltage created during load current interruption should be studied since the TIV is produced on top of the system voltage.

- The impact of EMI on the HVDC CB should also be investigated considering actual field operation conditions. Especially the state-of-the-art HVDC CBs have a lot of electronics on-board for control and actuation purposes. The robustness and immunity of these components against field operation environment need to be investigated.

- All HVDC CBs have mechanical switching devices operated by novel electromagnetic actuators. The mechanical performance of a multitude of novel high-speed actuators shall be investigated. Also a large number of vacuum gaps are used in series, which was not applied in the past. Many challenges arising from this including mechanical synchronicity, reliability and power supply need to be studied. Moreover, the DC voltage withstand of the mechanical switching gaps need more attention. This has an impact on the required redundancy in the designs of a (large) number of gaps in series. This requires investigation.

- Some HVDC CBs naturally recover after re-ignitions and restrikes while some do not. The impact of re-ignitions and restrikes on the overall operation of the MTDC grids needs to be investigated. This is essential for defining tolerance for pass/fail criteria during short-circuit testing of HVDC CBs.

- Hybrid HVDC CB technologies have PE components in the continuous current branch that conduct the load current during normal operation. The impact of long duration DC current conduction (without switching) on the PE switches shall be studied. This is essential input for specifying the requirements of pre-conditioning of this current branch during laboratory testing.

- Aggregate U-I characteristics of energy absorbing MOSA are nowadays pretty straight forward for simulation purposes. However, full stress analysis including thermal, mechanical, EM, dielectric, gas-discharge etc., processes are necessary and studies based on multi-physics approach are essential.

- Electrical and mechanical endurance tests of the whole HVDC CB reflecting the realistic application in MTDC grid needs to be defined.

A. Appendix

A.1. HVDC Switchgear

HVDC switchgear is used in every HVDC substation despite the type of the converter technology. Four major categories of HVDC switchgear have been identified depending on their functions as shown by different colors in Figure A.1 [Cig17a]. These are earthing switches (green), disconnect switches (blue), transfer switches (black) and CBs (red). Of all the HVDC switchgear the most notably missing in the HVDC substations is the HVDC CB.

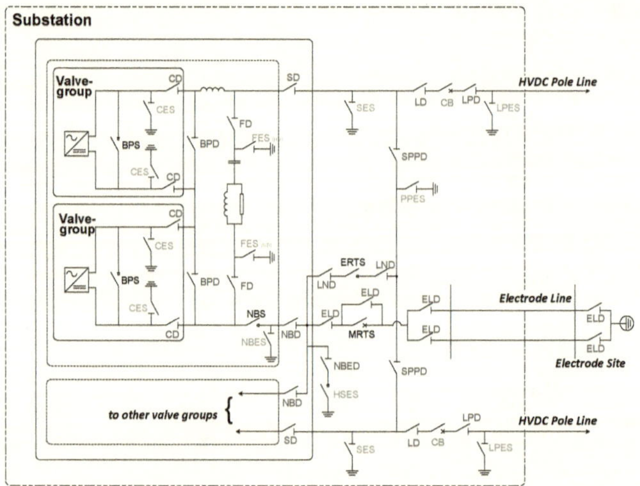

Figure A.1.: HVDC Switchgear layout in HVDC substations [Cig17a]

Also, of all the HVDC switchgear, the HVDC CB has the most onerous duty – clearing DC short-circuit current with very high urgency (within a few milliseconds). To achieve DC current interruption, HVDC CB need to rapidly create, maintain and withstand a counter voltage higher than the system voltage, which leads to system current suppression while absorbing the magnetic energy stored in the system. This is discussed in detail in Chapter 2

Similarly, testing of HVDC CB is also challenging because a test environment must reproduce the necessary stresses as in service condition. High-power DC sources are required for this purpose, which are currently not available at any test facility. However, testing of other HVDC switchgear does not require special test circuits nor procedures since standard test circuits available at test laboratories can be used.

Considering an HVDC CB, a type test program involves the following category of tests;

- Dielectric tests – High voltage withstand capability between terminals in open position or terminal-to-earth in a closed position. This includes short duration DC voltage withstand, partial discharge, lightning and switching impulse tests. It also includes high-voltage withstand tests of support structures and any components such as high-voltage power supply, communication links and cooling that is connected between high-voltage potential and earth.

- Operational tests – Capability of an HVDC CB to withstand any of the stresses resulting from (including worst-case) normal operation. This includes measurement of resistance, short-time current withstand, temperature rise, etc., tests.

- Making and breaking tests – This is intended to verify the performance of an HVDC CB when making and breaking of short-circuit currents.

- Endurance tests – This includes mechanical as well as electrical endurance of the HVDC CB.

All of the above tests except the making and breaking tests can be performed using standard test circuits and test procedures. But making and breaking tests are challenging because these tests require special test circuits capable of supplying the specified test parameters. For this reason this book focuses on the latter.

A.2. Definitions and Terminology Associated with the DC Interruption Process

Although the actual operation of various HVDC CB technologies varies, the general interruption process is similar – creation of a counter-voltage that exceeds the system

voltage and absorption of system inductive energy. As such, it is possible to define
a common terminology regarding the operational modes and timings. CIGRÉ JWG
A3/B4.34 ([Cig17a]) developed terminology associated with voltage and current waveform,
and timing definitions related to fault current interruption by a generic HVDC CB.
The most important timing and wavetrace definitions are described with reference to
Figure A.2 [BPS18b].

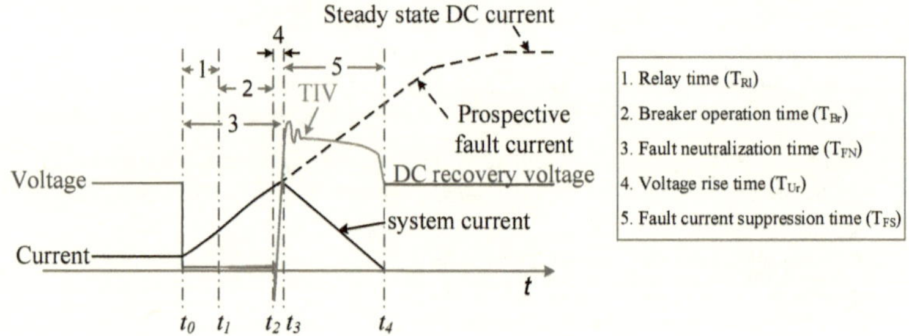

Figure A.2.: Diagram for illustration of timing definitions and terminology related to
HVDC CB operation. ©[2018]IEEE.

A.2.1. Timing related terminology

Relay time (T_{RI}) – the time interval between fault inception and the sending of a trip order
to the HVDC CB, (from $t_0 - t_1$, in Figure A.2). Although this is a characteristic of
the protection system and is not part of the test requirements of the HVDC CB, it
has a significant impact on the design (requirements). In fact, during a laboratory
test, a trip signal is fully controllable and is sent by the test circuit. If the HVDC
CB has self-protection functionality, this can be tested separately.

Breaker operation time (T_{Br}) – the time interval between the reception of the trip order
and the beginning of the rise of the TIV, (from $t_1 - t_2$, in Figure A.2). This
determines the speed of operation of the HVDC CB, which is a critical parameter
for the application in the envisaged MTDC grids. Thus, it must be clearly noted
during tests.

Fault neutralization time (T_{FN}) – the time interval between fault inception and the instant when the fault current starts to decrease, (from $t_0 - t_3$, in Figure A.2). This is important from system perspective since at the end of this time the system voltage starts to recover.

Voltage rise time (T_{Ur}) – the time required to build the TIV, (from $t_2 - t_3$, in Figure A.2). The rate-of-rise of the TIV is important because the dielectric recovery of the mechanical switching device(s) must be coordinated with this.

Fault current suppression time (T_{FS}) – the time interval between the peak of the interrupted fault current and the instant when the current has been suppressed to zero (arrester leakage) current level, (from $t_3 - t_4$, in Figure A.2). This time depends on several system and CB related parameters. From the CB perspective, two events take place during this period. The HVDC CB absorbs magnetic energy stored in the system's inductance and the TIV sustained by the MOSA appears across the HVDC CB during this period. From the system perspective, the voltage recovers during this period.

Break time – the duration from the trip order until the fault current is reduced to the residual current level.

A.2.2. Wave trace related terminology

Prospective fault current – the fault current that results at a CB location when no action to clear the fault is made. The rate-of-rise and the peak value of current are essential parameters.

Peak interruption current – the maximum value of the fault current that flows during the current interruption process. The peak value of the interruption current occurs at the moment the TIV equals the system voltage.

Transient Interruption Voltage (TIV) – the voltage that the HVDC CB generates[1] during current interruption, see the red trace in Figure A.2. During a test, the HVDC CB must clearly demonstrate that this voltage is higher than the rated operation voltage of the CB with sufficient margin. 50 % margin has been suggested as typical compromise between system insulation coordination and CB performance. The

---

[1]In fact, the HVDC CB does not have an active voltage source that generates a TIV, rather it is a sheer action and provision of the HVDC CB itself combined with the system inductance that leads to the generation of this voltage – see Section 2.2.

initial value, the rate-of-rise, the peak value and the duration of the TIV are crucial parameters.

Rated short-circuit breaking current – The maximum current that the HVDC CB can interrupt within specified breaker operation time.

Leakage (residual) current – the current that flows through the MOSA after fault current interruption.

The above and other terminology are described in detail in [Cig17a]. Interruption of load/nominal current follows the same procedure except in this case the relay time is not critical since the trip order can be directly sent to the appropriate CB. This can be performed, for example, for maintenance and/or reconfiguration purposes without de-energizing the converters or the other DC side of the system.

A.3. CIGRÉ B4.57 Benchmark Grid

Acknowledging the importance of having a unified benchmark system for harmonizing the research related to MTDC grids, Cigré WG B4-57 has developed an 11-terminal HVDC system shown in Figure A.3 [Cig14]. The system consists of three DC subsystems – DCS1 (top light blue – Bm-A1 and Bm-C1), DCS2 (bottom light blue – Bm-B2, Bm-B3, Bm-B5, Bm-F1 and Bm-E1) and DCS3 (dark blue – Bb-A1, Bb-C2, Bb-D1, Bb-E1, Bb-B1, Bb-B4 and Bb-B2).

The DC grid integrates wind power generations, offshore load, inland generation and in land loads (in the AC system). The DC grid has two types of transmission voltages: ±200 kV with symmetrical monopole converters (DCS1 and DCS2) and ±400 kV with bipolar converters (DCS3). A DC/DC converter (Cd-E1) connects DCS2 and DCS3 (200/400 kV). Furthermore, it consists of AC (380 kV onshore and 145 kV offshore) and DC grids having multiple in-feeds with AC and DC overhead lines as well as AC and DC cables (underground/submarine). Detailed data and control parameters for steady state power flow have been provided. In some studies only a part of the Cigré benchmark network, especially DCS3, is used as a test network [WRL14].

A.4. Additional Information on Developments of HVDC CBs

In order to get a visual impression, photos of HVDC CBs recently installed in HVDC converter stations (in China) are shown in Figure A.4 and A.5. Figure A.4 shows two in-service hybrid HVDC CBs (of full-bridge IGBT based topology) in series with current

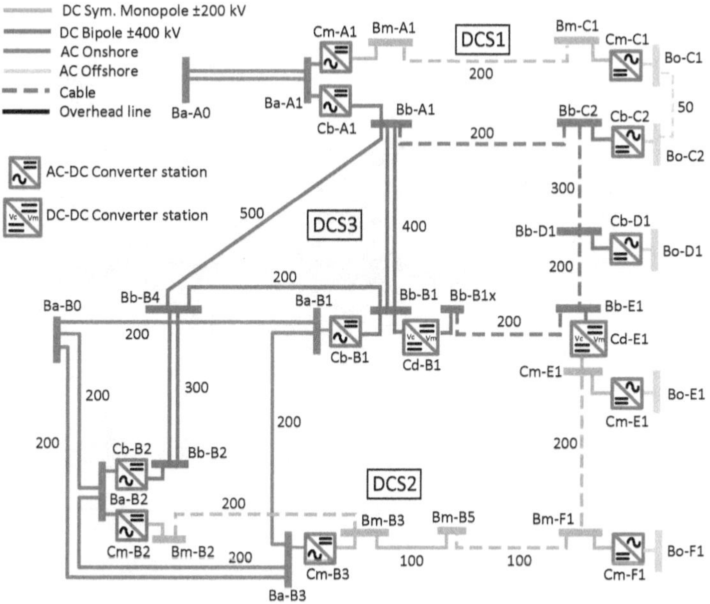

Figure A.3.: Cigré B4 DC Grid Test System according to [Cig14]. Numerical figures on the lines show line length in km, symbols designate as follows: "Ba" onshore AC bus, "Bo" offshore AC bus, "Bm" symmetric monopole DC bus, "Bb" bipole DC bus, "Cm" monopole AC-DC converter station, "Cb" bipole AC-DC converter station, "Cd" DC-DC converter station

limiting reactors installed at the ±200 kV Zhoushan five terminal HVDC pilot project [THP+15, TZH+]. In their CCB, these breakers consist of six series connected fast vacuum mechanical switches, which must achieve sufficient dielectric distance to withstand the TIV (of 300 kV within 2 ms) before breaking the current in the commutation branch [WLL+19]. Nevertheless, the vacuum switches in the hybrid HVDC CBs open under no current condition - thus arc-less operation is ensured since current is first commutated to the commutation branch by the PE switches in the CCB.

Figure A.4.: Hybrid HVDC CB – ±200 Zhoushan Five-Terminal MTDC grid [TZH⁺, TWZ⁺]

Figure A.5a and A.5b show the front and rear views of one of the two active current injection HVDC CBs installed at a converter station of Zhangbei ±500 kV MTDC pilot project. This breaker uses PE making switch instead of mechanical HSMSs as discussed in Section 3.3.2 [GFG19]. In the CCB, it consists of 12 VIs, which are connected in series to sustain a TIV of over 800 kV within 3 ms after a trip command.

A 500 kV realization of the optimized hybrid HVDC CB (modified topology discussed in Section 3.4.2) is developed for application in the Zhangbei MTDC grid [TZH⁺]. Figure A.5c shows a photo of one of the twelve hybrid HVDC CBs installed at one of the converter stations of the Zhangbei project. The CCS is composed of 14 series × 6 parallel PressPack IGBTs with a commutation time less than 200 μs [TZH⁺]. A UFD is composed of 10 series connected vacuum switches each of which are operated simultaneously by electromagnetic repulsion mechanisms. It is reported that within 2 ms these vacuum switches can achieve a total insulation level of 1000 kV with a voltage unbalance of less than 5%. The commutation branch is divided into modules (see Figure 3.10) where each module is a full-rectifier diode bridge consisting of 32 series connected IGBTs and 32

(a) Active current injection – front view

(b) Active current injection – rear view

(c) Hybrid – modified topology

(d) Hybrid – NVC topology

Figure A.5.: HVDC CBs installed at ±500 kV Zhangbei MTDC project. Courtesy: (a) and (b) Sieyuan Electric Company, (c) NR Electric, (d) C-EPRI

series connected diodes in a valve. 10 such modules are connected in series for a 500 kV rated voltage. Each module has a 15 MJ energy absorber MOSA at clamping voltage of 80 kV. This makes the total energy absorption capability of the entire hybrid HVDC CB 150 MJ. Each MOSA module has 40 parallel columns where each column is composed of 12 series stacked MO varistor blocks.

Figure A.5 shows another hybrid HVDC CB installed at one of the Zhangbei converter stations. This hybrid HVDC CB is based on negative voltage coupled topology – two realizations of which are installed in the project. It consists of eight 100 kV FMSs and 320 bidirectional SMs in the commutation branch [ZZXS20]. The SMs in the commutation branch are divided into five breaker units each with common MOSA branch. The FMSs are actually commutating the current from the CCB when the current zero is created. It cannot be considered as interruption since shortly after current commutation there is no voltage seen by the FMSs. This is because the commutation branch keeps conducting (with no charging of any capacitor) for a few milliseconds until the FMSs gain sufficient dielectric strength – a completely different stress than in the mechanical active current injection HVDC CBs.

The other critical aspect making the testing of HVDC CBs challenging is the size and footprint of these equipment. Not only the supplying the rated test parameters but also fitting the fully rated devices into the test environment and the associated logistics become challenging. For example, the HVDC CBs developed for Zhangbei project, shown in Figure A.5, are reported to have the following tower dimension: 18 m×15.5 m×9 m (W×H×D) and weight of nearly 180 t [Cao].

A.5. Impact of Traveling Waves on Insulation Coordination

An important aspect of traveling waves during fault current interruption is its impact not only on the rate-of-rise of fault current but also on the insulation coordination, specifically, on terminal-to-ground withstand voltage of an HVDC CB and also on the components connected thereby. It was discussed (Chapters 4 and 5) that a fault occurring close to a converter terminal presents the worst-case electrical stresses to the HVDC (in terms of current, energy and voltage between terminals) because it results in the highest average rate-of-rise fault current considering the achievable fault neutralization time T_{FS}. However, when a fault occurs far away from a converter terminal, the impact of the traveling waves become prominent. Figure A.6 shows fault current interruption at 240 km from a converter initially receiving power. The dashed curves show prospective current and terminal-voltage on the cable side of the current limiting reactor L_{DC}.

The converter side voltage of the L_{DC} is shown by solid blue line in Figure A.6. During

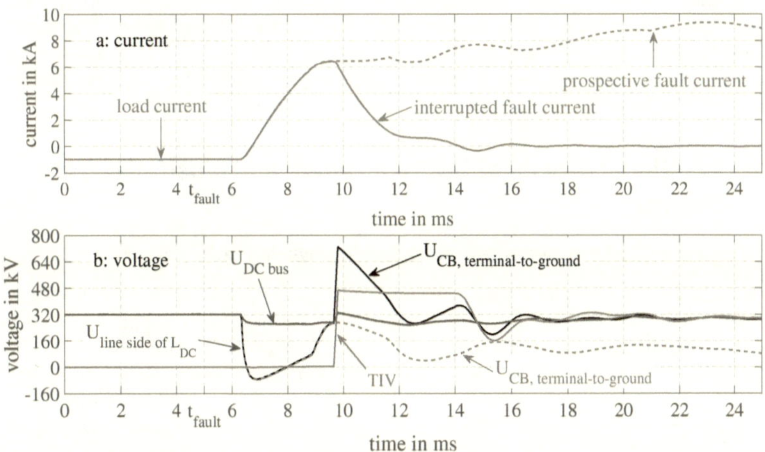

Figure A.6.: Simulation result showing impact of traveling waves on the terminal to ground voltage of HVDC CB

fault current interruption process, the TIV produced by the HVDC CB superimposes on to the oscillating traveling waves. Depending on the position of the L_{DC} relative to the HVDC CB and/or interruption current direction, the voltage-to-ground at the CB terminal and can (theoretically in case of zero attenuation) reach as high as 2.5 p.u. value when a fault occurs far from the CB. However, practically surge arresters may be installed at this location to protect against such overvoltage. Nevertheless, the insulation coordination of the line surge arresters in the presence of HVDC CBs needs further investigation.

A similar situation, although depending on the HVDC CB technology, can be observed during a load current interruption. During a DC load current breaking, the TIV produced by the breaker is superimposed on to the system voltage since the transmission lines are charged to the system voltage unlike during a fault. This requires due attention when designing the insulation coordination of both the system as well as the HVDC CBs.

Although the discussion in this section was not included in the test requirements defined in this **book**, this must be considered for the complete type test program. This is important because it verifies the high-voltage withstand of support structure and any components such as high-voltage power supply, communication links and cooling that is connected between high-voltage potential and earth. On the other hand, this consideration contributes significantly to the overall size and footprint of HVDC CB, thus creating additional challenges to the test environment to accommodate the complete CB.

A.6. Additional Details of the Test Circuit and Illustrative Test Results

A.6.1. Test circuit implementation for testing hybrid HVDC CB

Depending on the type and ratings of the HVDC CB to be tested, some modifications to the test circuit discussed in Section 8.2 is needed. For example, compared to the test circuit shown in Figure 8.2, the test circuit used for the demonstration of the 350 kV hybrid HVDC CB has been slightly modified and rearranged in order to take into account the TIV which is in the range of 500 kV. These can be seen in the electrical diagram of the test setup shown Figure A.7. Figure A.8 depicts the actual test setup and components of the test circuit in a test laboratory. There are three main modifications:

- Two series connected spark gaps are used for the overcurrent protection instead of one TSG_1.

- AB_1 and AB_2 are, respectively, replaced by double interrupting chamber HVAC CBs (rated for 420 kV, 50 kA AC) instead of single interrupting chamber HVAC CBs used in other demonstrations.

- The arcing prolongation circuit has been relocated on grounded side of the DUT in order to avoid excessive terminal-to-ground voltage across this sub-circuit. Also additional overvoltage protection MOSA module has been put across the terminals of the arcing prolongation circuit to avoid damage to the current limiting reactors.

As the rating of the HVDC CB increases, the minimum required gap length of the TSG_1 increases due to the increase in the TIV. This poses a challenge in the triggering of the spark gap since the bridging distance of the triggering plasma is limited. In order to trigger a spark gap, the plasma ejections from either side of the gap must reach one other to create a short-circuit. For example, during the testing of the 350 kV hybrid HVDC CB, the gap length was initially set to 45 cm to avoid spontaneous breakdown due to the resulting TIV. However, this gap length was found to be too long to establish a short-circuit path for the ejected plasma. In order to mitigate this, two triggered spark gaps are used in series as shown in the Figure A.7 and its implementation is shown in Figure A.8c. Then the gap length of each the spark gap is set to 15 cm. The total dielectric distance is shorter (30 cm) in the latter case because the spontaneous breakdown distance of two spark gaps in series is higher than in only one spark gap of the same overall gap distance. Also larger diameter (60 cm) TSGs are used instead of a smaller diameter (30 cm) TSGs[2]. This helps in better electric field distribution to avoid unintended dielectric breakdown of the gaps.

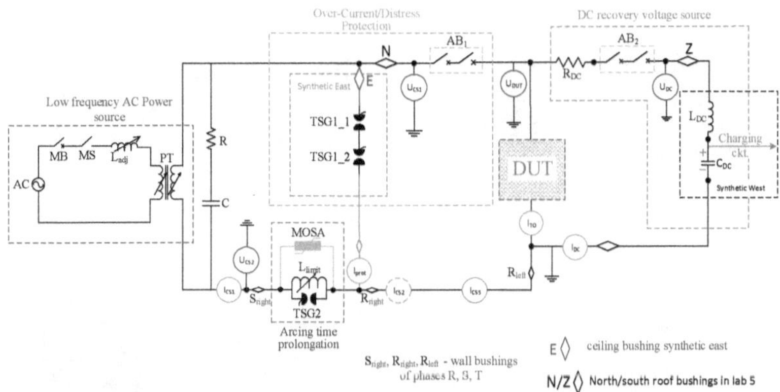

Figure A.7.: Electrical diagram of the test circuit used for 350 kV hybrid HVDC CB

The other challenge is related to the insulation coordination of the arcing time prolongation circuit also due to the increase in the TIV. Actually, there are two issues that need to be addressed here. The first is insulation from the ground of this sub-circuit. This is because the TIV of the DUT appears across terminal-to-ground of both the TSG_2 and L_{limit}. The second issue is the risk of high-current flow through L_{limit} especially when the TSG_2 does not operate for any reason. The latter is due to the fact that the peak value of the source voltage applied during a test increases with the rating of the DUT. Thus, a large asymmetric current can result in L_{limit} if the TSG_2 does not trigger. This can result in damage to these reactors if they are not rated for high-current withstand especially during prospective current calibration tests[3] and/or when the DUT fails to clear while

---

[2]The latter is used in the arcing prolongation circuit

[3]At this step the DUT does not receieve a trip command and hence, remains in a closed position or if

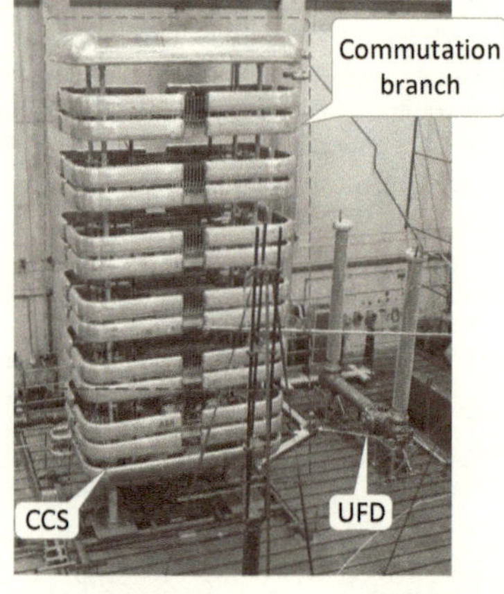

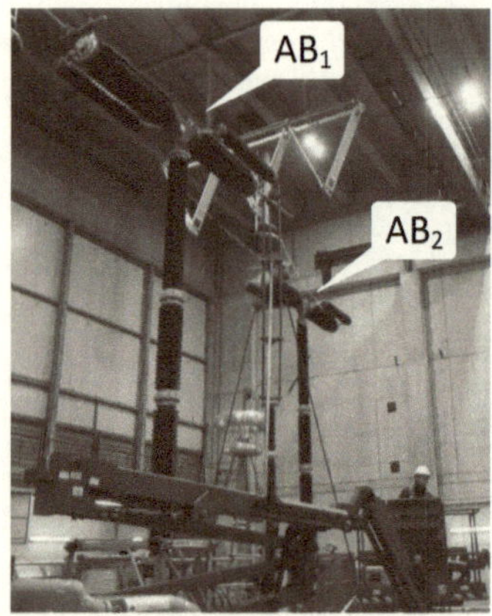

(a) 350 kV Hybrid HVDC CB (DUT)   (b) Two double chamber interrupters ABs

(c) Two series connected TSGs (d) Current limiting reactors of arcing prolongation circuit

Figure A.8.: Actual test setup of 350 kV hybrid HVDC CB – DUT and major test circuit components are shown

---

there is a risk of large current flow in the DUT it can even be bypassed by a parallel short-circuit path. In the latter case the DUT is set to remain in open position at the same time the TSG_2 fails to trigger – particularly the force between the winding turns becomes the main issue.

Therefore, in the designed test circuit, some practical risk mitigation steps have been taken when testing higher rated HVDC CBs. To avoid the impact of the high terminal-to-ground voltage on the arcing prolongation circuit, the latter is moved to the grounded (return current) side of the test setup – see Figure A.7. In this case the arcing prolongation circuit does not face the TIV stress. However, proper care must be taken to avoid damages to the devices (including measurement equipment) connected to the return line, which are located on the source side of the arcing prolongation circuit, since the voltage drop across the reactor when current flow during current limiting mode can be significant and this voltage is "seen" by these equipment.

The risk of large current through the current limiting reactor is mitigated by using multiple reactors in parallel so that current is shared in case the TSG_2 fails to trigger. Thus, large reactors are used to obtain the desired reactance when combining in parallel. For example, four current limiting reactors L_{limit} each with 162 mH[4] of inductance are used in the arcing prolongation circuit, the actual implementation of which is shown in Figure A.8d. In this case two sets of reactors are connected in parallel where each set consists of two series connected reactors. An equivalent reactor of 162 mH inductance is achieved as a result. In addition, to protect the arcing prolongation reactors from conducting large currents, a MOSA module is placed in parallel with the arcing prolongation circuit – see Figure A.7. The MOSA module starts to conduct when the voltage across the arc prolongation reactor reaches a certain threshold (for example, 35 kV), which occurs only when the reactors are conducting large currents. Besides, the gap distance of the TSG_2 is set such that it will break down spontaneously if the voltage across the reactor reaches a value above the expected protection level of the MOSA module – thus providing double protection.

The photo of the DUT (the 350 kV hybrid HVDC CB) is shown in Figure A.8a. Figure A.9 depicts 16 kA (TF100*[5]) current breaking by the hybrid HVDC CB. The electrical diagram of the DUT is depicted in Figure A.11. This is a design based on BIGTs (Bi-mode Insulated Gate Transistors). After 3 ms from the trip command the CB started to produce the TIV which rises to nearly 490 kV. The current commutation from the CCB to the commutation branch is shown in the zoomed section of the graph as illustrated by different colors. The hybrid HVDC CB suppressed the short-circuit current within 2.5 ms, where 9.2 MJ of energy is absorbed within the same amount of time. The energy absorption is plotted in Figure A.10.

---

[4]each rated for maximum short-circuit current of 2000 A

[5]Later TF100* was increased to 20 kA

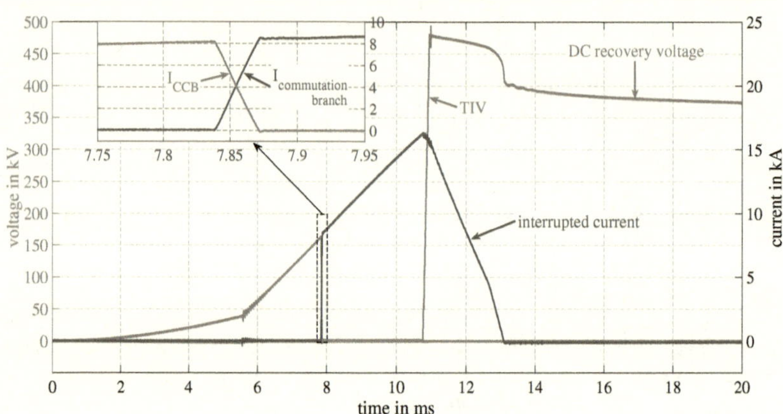

Figure A.9.: 16 kA Test result of hybrid HVDC CB - 350 kV, 16 kA – With DC recovery voltage application

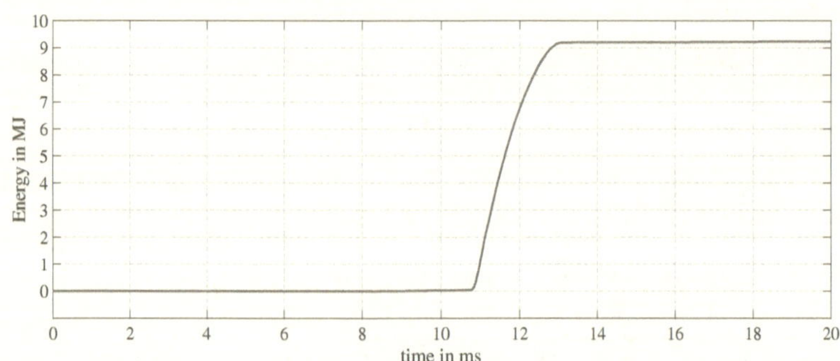

Figure A.10.: Energy absorption by hybrid HVDC CB during 16 kA current interruption

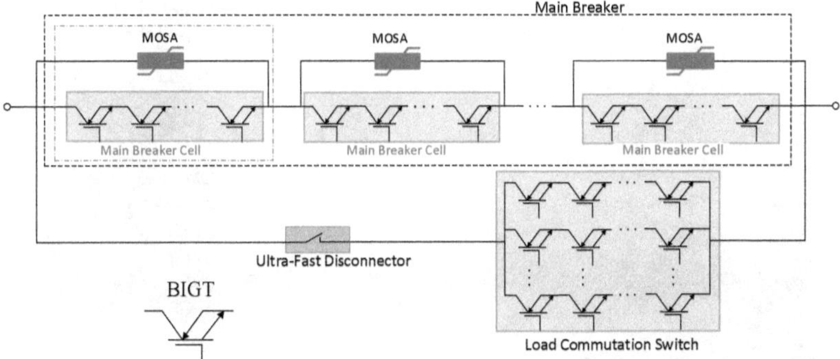

Figure A.11.: electrical layout of hybrid HVDC CB - original topology based BIGTs

A.6.2. Additional test results of 160/200 kV active current injection HVDC CB

A photo of the laboratory prototype installation of the active current injection HVDC CB tested at KEMA Labs is shown in Figure A.12. The prototype breaker has two series connected high-voltage VIs in the CCB. The counter current is injected from a pre-charged capacitor by closing two series connected mechanical HSMSs (also vacuum) [TIS+19]. The main components of the prototype breaker are labeled in the figure. The test results of this demonstration are discussed in Chapter 8 where 16 kA current interruption was achieved within 7 ms after trip command while producing a TIV with peak value of 250 kV. Figure A.13 shows an artist impression drawing of a compact 320 kV active current injection HVDC CB. Four VIs and four vacuum HSMSs are used in this case.

Figure A.14 shows a test result of the active current injection HVDC CB discussed in subsection 8.3.3. The same breaker tested for 160 kV system voltage was tested for 200 kV system voltage by modifying the MOSA where the clamping voltage is increased by adding a series MOSA module. To reflect this, the counter current injection capacitor bank is pre-charged to 200 kV, and additional reactors are installed to increase the inductance of the injection circuit in order to maintain the peak value of the injection current as in 160 kV case. The breaker interrupts current of 17.2 kA while producing the TIV with peak value of nearly 350 kV.

Test circuit parameters are also adjusted to take the new breaker parameters into account. Particularly, the source voltage and circuit inductance are increased to keep the

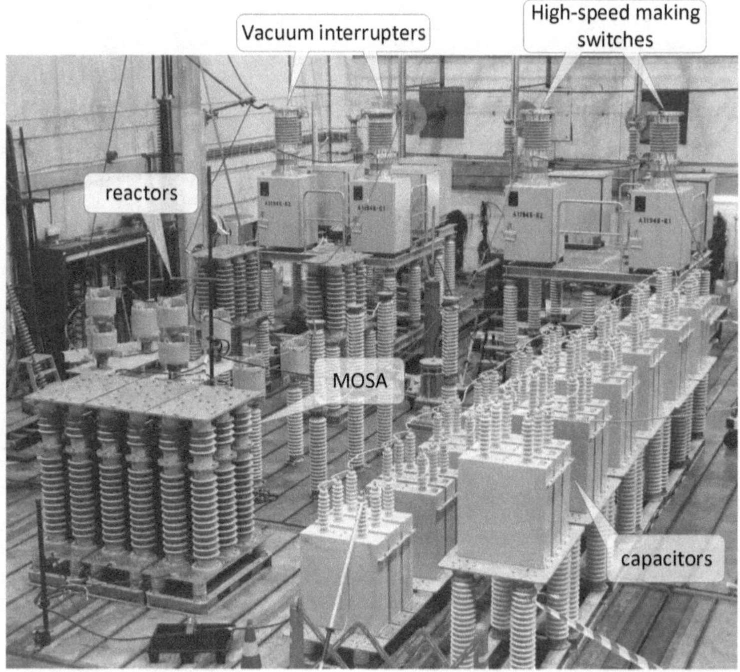

Figure A.12.: Active current injection HVDC CB – 160/200 kV prototype tested at KEMA Labs. [by kind permission of MITSUBISHI Electric Europe]

current suppression period as in 160 kV test – resulting in a higher energy absorption of nearly 4 MJ in this case. It can be seen from Figure A.14 that the TIV remains above 300 kV during the entire current suppression period, which lasted for about 1.75 ms. After the current suppression the DC recovery voltage of 230 kV (assuming 15% continuous operation overvoltage) is applied (at 22nd ms on the graph) for about 1 s. The current suppression is completed around 18.6 ms in the figure. Between 18.6 ms and 22 ms a slightly decaying (due to conduction through the EAB) self-imposed DC recover voltage,

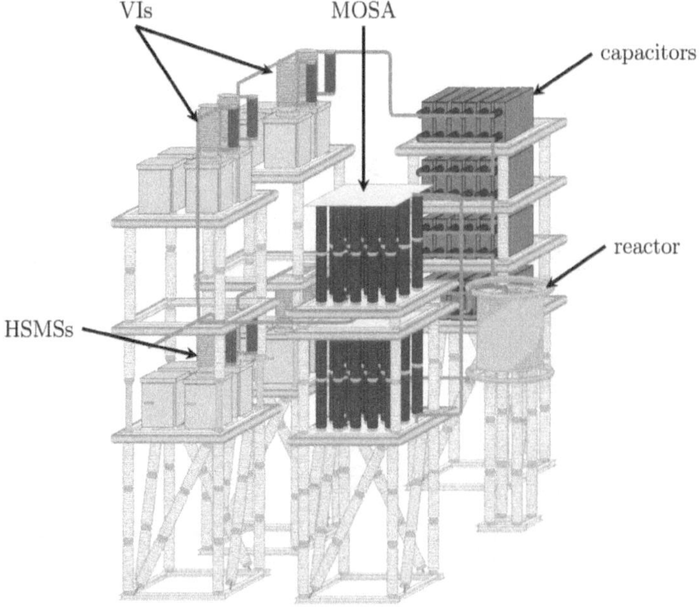

Figure A.13.: Rendering of 320 kV active current injection HVDC CB (courtesy Mitsubishi Electric)

which is higher than the DC recovery voltage supplied by the test circuit is seen on the graph. This is due to the charge across the capacitor bank of the DUT that is trapped at the end of the current suppression period. This capacitor remains charged to the same voltage level as the TIV of the breaker during the current suppression period. This could be an ideal source of DC recovery voltage but its continuity depends on internal actions of the CB. Also, not every design of HVDC CBs exhibit the same behavior. For example, if PE making switches are used for injecting counter current, they can be turned off (actively or naturaly as in the case of thyristors) to prevent the voltage stress coming from this capacitor on other internal components of the HVDC CBs.

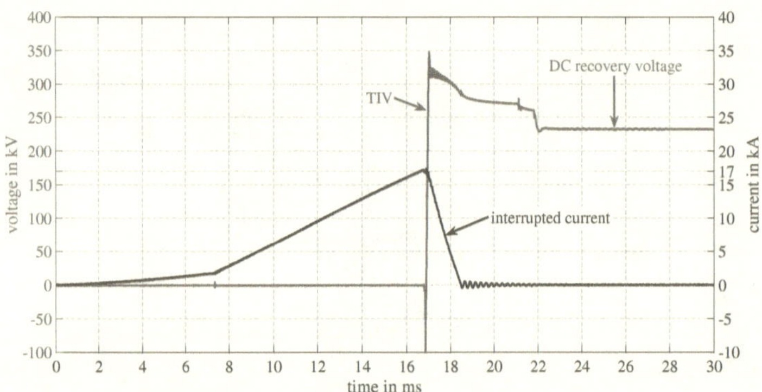

Figure A.14.: A test result 200 kV active current injection HVDC CB, 16 kA (TF100*) current interruption

A.6.3. Test results of VARC HVDC CB

Like all other mechanical HVDC CBs, the VARC HVDC CB is naturally bidirectional. Its electrical diagram and the detailed operation principle is discussed in Section 3.3.4. Within the framework of PROMOTioN project, two prototype realizations of this concept, at different ratings, have been developed and tested at KEMA Labs.

A photo of the first single module VARC DC CB prototype tested at KEMA Labs in June 2018 is shown in Figure A.15a. The single module prototype was designed to produce and withstand a 40 kV TIV while breaking 10 kA. The EAB for the module was designed to dissipate 2.5 MJ energy before cooling to ambient temperature. The prototype could achieve up to 11 kA current interruption within the breaker operation time of 3 ms while producing a TIV of sligthly higher than 40 kV and energy absorption of nearly 1 MJ. The TIV of one module corresponds to a 27 kV system voltage. This means 12 modules would be required for a 320 kV breaker [ANM+19, LPM+20]. The artist impression rendering of a 320 kV design is shown in Figure A.15b together with a residual current breaker and a DC current limiting reactor as labeled in the figure.

In august 2020, an 80 kV, 15 kA VARC HVDC CB consisting of three independent modules connected in series has been tested. The photo of the test setup, including some components of the test circuit, is shown in Figure A.16. An important feature in this modular design is that the modules of the CB operate independently from each other. In

(a) Photo of 27 kV VARC DC CB Tested at KEMA Labs' high-power laboratory

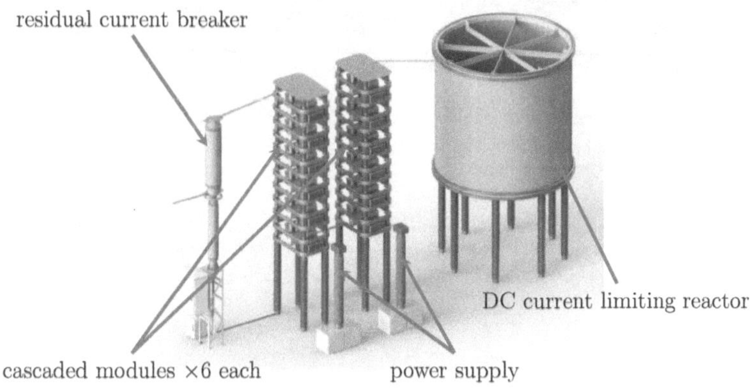

(b) Artist impression (rendering) of a 320 kV VARC HVDC CB (courtesy: SciBreak AB)

Figure A.15.: Designs of VARC HVDC CB

this case the breaker operation time of <2 ms and a moderate energy absorption capacity of 3 MJ (1 MJ per module) has been demonstrated .

Figure A.16.: Test setup of three module 80 kV VARC HVDC CB at KEMA Labs

All the test requirements defined in Chapter 8 have been fulfilled with bidirectional current interruption. Initially, TF100 of 12 kA was defined and this was successfully achieved within the breaker operation time of 1.5 ms. Later the TF100 is increased to 15 kA and this could be achieved within the breaker operation time of 2 ms. In both cases the breaker produced a TIV with peak value of 130 kV while up to 3 MJ energy absorption is demonstrated. The TF100* of 15 kA current interruption was performed for each current direction. Figure A.17 shows TF100* forward current interruption in which 15 kA is interrupted within the breaker operation time of 2 ms. The prospective current with a rate-of-rise of nearly 3.5 kA/ms is superimposed for comparison. In this case, the short-circuit current was suppressed within 2.2 ms while absorbing 1.8 MJ of energy. However, DC recovery voltage was not applied from the test circuit during the demonstration of VARC HVDC CB. The decaying DC voltage that is seen after current suppression in the figures is due to the trapped charges across the resonant capacitor which the breaker automatically discharges after current suppression is completed.

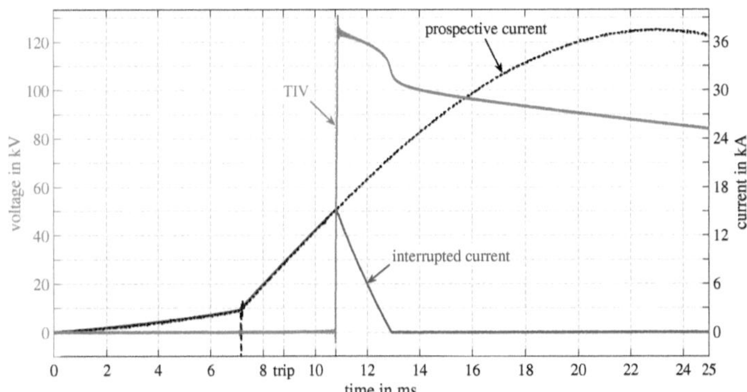

Figure A.17.: Demonstration of TF100* test of an 80 kV VSC assisted resonant current (VARC) HVDC CB

Next, the TDT test is performed, the test result of which is shown in Figure A.18. The purpose of the TDT test is to maximize the TIV duration without adversely affecting its rate-of-rise and peak value of the TIV compared to its value during the rated current interruption. The TDT test of the VARC HVDC CB is performed with short-circuit current interruption of 8.2 kA. The test circuit parameters are adjusted so that, in this case, the current suppression duration of 5.3 ms is achieved with energy absorption of nearly 2.5 MJ. More than twice the duration of the TIV obtained during the TF100* test is achieved during the TDT test.

Finally, the TC100 current interruption (both in the forward and reverse directions) have been performed to verify the low-current interruption capability and performance of VARC HVDC CB. The TC100 corresponds to 2000 A continuous current and the oscillograms of the test result where reverse current is interrupted are shown in Figure A.19.

The VARC HVDC CB technology enables an intelligent control of the conditions near the current zero in the main interrupter. This control is achieved by the VSC onboard the breaker that excites the an increasing resonant current as discussed in Section 3.3.2. As a result, the $\mathrm{d}i/\mathrm{d}t$ at current zero and the ITIV following current interruption are not so severe compared to the values at high-current interruptions. This is because the current interruption is achieved when the resonant circuit capacitor is charged to a voltage

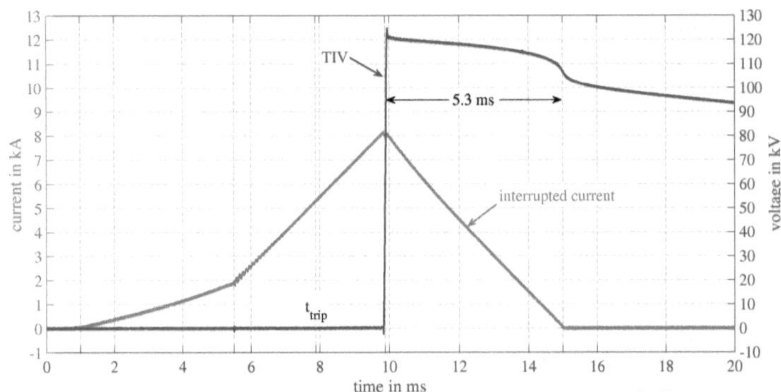

Figure A.18.: Demonstration of TDT test of an 80 kV VSC assisted resonant current (VARC) HVDC CB

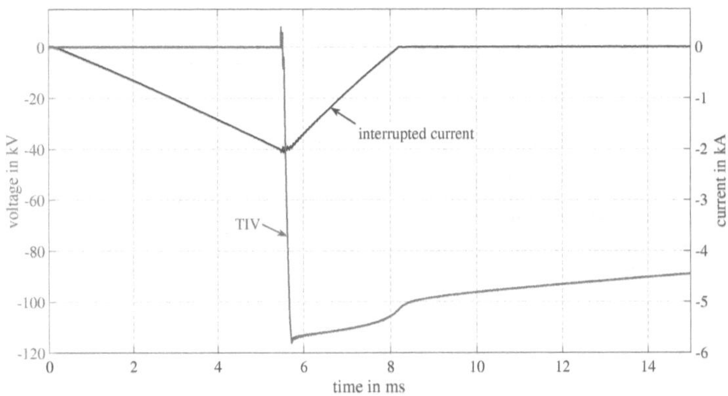

Figure A.19.: Demonstration of TC100 test of an 80 kV VSC assisted resonant current (VARC) HVDC CB

just enough to create current zero with not so significant remnant charge across it – see Figure A.19.

A.6.4. Additional information on transformer saturation

Figure A.20 depicts the primary side measurements (corresponding to the test results shown in Figure 8.19) where the current supplied by one of the four short-circuit generators is shown along with the generator terminal voltage. Note that the generator terminal voltage shown in Figure A.20 is not the same as the transformers' primary voltage due to the voltage drop across the adjustable reactor located between the two – see figure 8.2. Nevertheless, it can be seen that the primary side (generator) current is never suppressed to zero unlike the secondary (DUT) side current shown in Figure 8.19. Actually, after the transformers saturate, the rate of current suppression increases on the secondary side because of the significantly reduced circuit inductance seen by the TIV, whereas the rate of current suppression on the primary side remains unchanged. Hence, by the time the secondary current is suppressed to zero, considerable current may still flow through the primary winding. Without transformer saturation, the rate of decay of the primary and secondary currents are related by the transformer ratio.

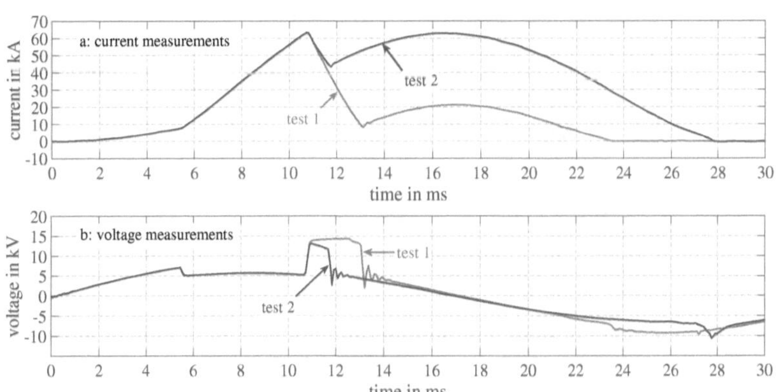

Figure A.20.: Impact of transformer saturation during current interruption by HVDC CB– comparison of primary (generator) side current (per generator) and voltage measurements at two successive tests

After the secondary side current is suppressed, the generator "sees" the saturated transformer as if its secondary side is still short-circuited. From this moment on, the primary winding faces large short-circuit currents although the secondary winding is actually an open circuit. Depending on the initial current and phase angle at the moment of secondary current suppression, very large currents ($>$60 kA per generator) might result. At the next current zero, the saturated transformer current ceases to flow since at that point the transformers momentarily come out of saturation.

A.6.5. Additional results of experimental DC CB

Figure A.21 shows a re-ignition that occurred during low-current interruption. The re-ignition occurred 0.45 ms after current interruption at the 1st CZC. This is one of the result obtained during the experimental investigation of VI type B discussed in Section 9.3.2. Several CZCs have been created until finally clearing at 22nd zero crossing.

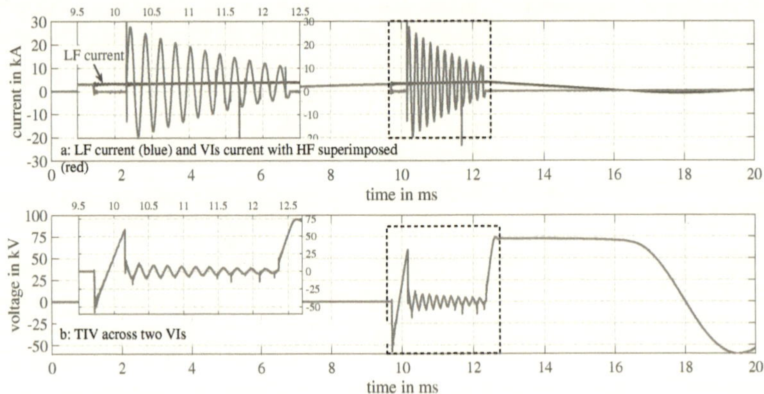

Figure A.21.: Low current interruption by double-break interrupter of VI type B after restrike

The main lesson from the test result of Figure A.21 is that low current interruption is not necessarily an easy task for HVDC CBs. It is not the intention of this book to judge whether re-ignitions and/or restrikes are acceptable, or whether the pass or fail criteria should be based on such phenomena. This is beyond the scope of this book, and one has to study the impact on these phenomena on the system to reach on fair judgment. However, it is the objective of this book to showcase the criticality of such a test and to explain the importance of including such tests in the test requirements.

A.7. MOSA V-I Characteristics used in the simulations

The aggregate V-I characteristics of the MOSA used in the simulation is shown in Table A.1. The voltage is per unitized with respect to the rated system voltage of the breaker being simulated. In PSCAD the actual value of the rated system voltage is provided to the model of a breaker.

Table A.1.: Aggregate U-I characteristics of MOSA used in the simulations

Current (kA)	Voltage (p.u.)
1.5E-6	0.17
4.5E-6	0.83
15.0E-6	1.16
75.0E-6	1.29
1.5E-3	1.33
15.0E-3	1.40
1.5E+0	1.47
18.8E+0	1.64
37.5E+0	1.69
75.0E+0	1.74
150.0E+0	1.84
225.0E+0	1.91
450.0E+0	2.01
750.0E+0	2.11
1.5E+3	2.26
3.0E+3	2.48
6.0E+3	2.79
15.0E+3	3.49